Hard Science,
Hard Choices

Hard Science, Hard Choices

FACTS, ETHICS, AND POLICIES GUIDING
BRAIN SCIENCE TODAY

Sandra J. Ackerman

DANA
PRESS

New York • Washington, D.C.

Published by Dana Press
New York/Washington, D.C.

DANA
PRESS

The Dana Foundation
745 Fifth Avenue, Suite 900
New York, NY 10151

900 15th Street NW
Washington, DC 20005

DANA is a federally registered trademark.

ISBN-13: 978-1-932594-02-7
ISBN-10: 1-932594-02-7

LIBRARY OF CONGRESS CATALOGING-IN-PUBLICATION DATA
Ackerman, Sandra.
        Hard science, hard choices : facts, ethics, and policies guiding
    brain science today / by Sandra J. Ackerman.
            p. 153
    Includes bibliographical references and index.
    ISBN 1-932594-02-7
    1. Neurosciences.   2. Brain.   I. Title.
    QP376.A233 2006
    612.8—dc22
                                        2006002976
Cover design by Kristine Pauls
Text design by Kachergis Book Design

Contents

Foreword

by Ruth Fischbach and Gerald Fischbach

This book, *Hard Science, Hard Choices,* synthesizes for general readers a significant scholarly meeting called to discuss recent advances in neuroscience that give rise to ethical issues unprecedented in their consequence to individuals, society, and the body politic.

The Library of Congress invited the organizers to hold the conference in the Library's classical building on Capitol Hill. There, in May 2005, more than sixty scholars from the fields of neuroscience, law, public policy, and philosophy gathered. The two-day meeting was co-sponsored by the Columbia University College of Physicians and Surgeons, the National Institute of Mental Health, and the Dana Foundation, along with the Library.

The debate was intense, thoughtful, and rigorous. The subject, all participants agreed, is critically important. The brain, after all, is the seat of what we consider our humanity. Our growing understanding of how the brain works and how we may manipulate, inquire into, or change it (both to treat its disorders and for nonmedical purposes) must now call forth our best efforts to seek ethical consensus while issues are taking shape—not after they have emerged as moral crises or controversies in the public arena.

Two public events that bracketed the conference underscored the real-world context of the discussion. The first was the national trauma of the Terri Schiavo tragedy, illustrating so well where science, policy, and ethics intertwine. The nation received numerous lessons in neurology, and passionate public debate was conducted on several levels in the news and in Congress. The second public event actually brought the conference to an abrupt end just as the meeting's closing remarks were under way: The Capitol and surrounding buildings, including the Library, were evacuated as Air Force fighter jets scrambled to intercept a plane that had entered forbidden airspace over the city. Had the intrusion been more than the lost pilot it turned out to be, the debate over detecting suspicious behavior, interrogation techniques, and protecting troops fighting a deadly enemy would have returned to its anguished post-9/11 level. It is a debate with a quiet but extensive neuroscience involvement, as a speaker at the meeting had discussed only an hour before the evacuation.

The name of the field that has developed around public issues arising from brain science is *neuroethics,* and it is a very young discipline: only three years before this meeting, more than 150 people met in San Francisco in a seminal conference titled "Neuroethics: Mapping the Field," organized by Stanford University and the University of California–San Francisco. That conference had its genesis about eighteen months earlier, in a conversation between the Dana Foundation's then board member and now chairman, William Safire, and Dr. Zach Hall, then chancellor of UCSF and now director of the State of California's stem-cell research initiative. Hall was showing the Dana official around UCSF's new high-tech science facilities when their discussion turned to the possible misuse of the great power of insight into the brain that was becoming probable. Safire dubbed the potential problems "neuroethics," which to Hall seemed immediately to deserve exploration, and so the conference was planned.

The "Mapping the Field" conference and proceedings published afterward gave the field an important jump-start and galvanized interest in this area. The organizers of that conference defined neuro-ethics as "the study of the ethical, legal, and social questions that arise when scientific findings about the brain are carried into medical practice, legal interpretations, and health and social policy." With findings occurring in fields such as genetics, brain imaging, and disease diagnosis and prediction, the definition went on to say that neuroethics should examine how doctors, judges, lawyers, insurance executives, and policymakers, as well as the public, will deal with them.

But we found that William Safire's definition in later remarks at the conference cut right to the chase. He said, "Neuroethics is the examination of what is right and wrong, good and bad about the treatment of, perfection of, and welcome invasion or worrisome manipulation of the human brain."

Three years later, with the technology advancing inexorably, we gathered at the Library of Congress for the conference that we titled "Hard Science, Hard Choices: Facts, Ethics, and Policies Guiding Brain Science Today." Our intent was for the meeting to establish another clear milestone to mark the progress of the field and serve as a new, tangible reference point as these increasingly complex and fascinating issues continue to develop. We ourselves have spent our entire careers in ethics and neuroscience; the two subjects are not just careers, but our lifelong interest. Nonetheless, it is almost hard for us to believe the breathtaking advances in technology, and we have an increasing sense that the critical social, legal, economic, policy, and especially ethical implications resulting from these advances need to be examined.

The goals for the conference included an exploration of new and emerging technologies, exploration of ethical, social, economic, and legal implications of new technologies, implications for public policy, and to facilitate scholarly networking, a key element in any emerging field.

One of our most difficult decisions was to decide which amongst the myriad compelling topics would be our focus. Casting too wide a net would cost the meeting traction; discussion would have to be shallow and too much precious time would be spent on issues that today are only, in the biblical expression, "a cloud on the horizon, no bigger than a man's hand." After much consideration and (often reluctant) ruling out of areas, we settled on three that are not just tomorrow's issues but confront our society today: neuroimaging; robotics and other computer-brain neurotechnologies; and psychopharmacology, including enhancement. These are such skimpy words relative to the enormity of their immediacy and scope.

The overarching neuroethical concern is that, with all the hype as well as the hope, we must at the same time be circumspect about emergent technologies. Many advances may—and indeed undoubtedly will—have unintended consequences that we must manage. We are discovering insights into improving the disordered brain and at the same time advancing on how we can make good brains better. We should respect the technology for the benefits it offers, but limit its use when it tends to lead to harm. We may have to rein in the technological imperative that urges, "If the technology exists, use it." Even more, we need to recall the bioethics mantra: "It's not what you can do, rather it's what you should do."

In this period of heady development, we need to consider whether there are boundaries across which we should not tread. We need to consider who decides and on what basis a high-risk procedure is justified in a given individual. With enhancement techniques comes the need to confront questions of equity. We need to be especially concerned about those in a vulnerable or compromised position. How we protect the rights and safeguard the welfare of those who need our protection—such as, for example, those who volunteer themselves for research—reflects on how we as a society value and respect the vulnerable among us. And make no mistake: at some time in our lives, we will all be vulnerable.

When we deal with brain science, we are dealing with the organ that makes us unique individuals, that gives us our personality, memories, emotions, dreams, creative abilities, and at times our sinister selves. Although the Decade of the Brain has come and gone, we are yet to understand and know the mind. Now we are in the decade of behavior. Yet, despite all we are learning, the brain remains an enigma.

It was clear that the conference participants gained not only a greater appreciation of the complexities of the brain but also an increased admiration of those pushing the frontiers as investigators and as research participants. We concluded the conference aware that hard science demands hard choices and that good science goes best with good ethics.

Ruth Fischbach, PhD, MPE, and *Gerald Fischbach, MD,* conceived and organized the conference "Hard Science, Hard Choices" at the Library of Congress, May 10–11, 2005. Ruth Fischbach is professor of bioethics and director of the Center for Bioethics at Columbia University in New York City. She is a faculty member in the Department of Psychiatry at the Columbia University College of Physicians and Surgeons and in the Department of Sociomedical Sciences at the Columbia University Mailman School of Public Health. Gerald Fischbach is executive vice president for health and biomedical sciences and dean of the Faculty of Medicine at the Columbia University College of Physicians and Surgeons, where he is also Hatch Professor and professor of pharmacology in the Center for Neurobiology and Behavior.

Part One

Overview

Introduction ⟋

In 1864, one Señora Zentino of Cuzco, Peru, received a social call from American diplomat and scientist Ephraim George Squier. In the course of his visit, she showed him an astonishing archaeological object: a 400-year-old human skull bearing a square-shaped hole, which bore unmistakable signs of having been cut and having begun to heal while the individual was still alive. An expert on Peruvian antiquities, Squier instantly recognized the skull as an example of trepanation, or trephining, a surprisingly successful (and still occasionally used) surgical practice of removing a section of skull to enable recovery from certain head wounds. Although the practice of trepanation was known from ancient medical writings, this skull constituted the first solid evidence that the surgery had actually been performed and had even allowed the patient to survive for some weeks afterward. The following year the Cuzco skull caused a sensation at the New York Academy of Medicine, where it exemplified both the daring and the venerable history of medical approaches to the brain.

Advances in the healing arts have always been both driven and restrained by two questions. The first of these is, What can we do? The second is, What *should* we do? In the study and practice of medicine, the two questions have been yoked together since at least the time of Hippocrates, whose timeless oath for the swearing-in of new doctors contains the injunction, "First, do no harm."

In recent years, the area of medical ethics that applies specifically to the study and treatment of the brain has drawn in thinkers from fields ranging from philosophy and psychiatry to law, psychology, and public policy. A conference sponsored by the Dana Foundation in 2002 brought together more than 150 professionals, including many eminent neuroscientists, under a new banner: neuroethics. The participants charged themselves with studying the ethical questions that arise when scientific findings about the brain are carried into medical practice, legal interpretations, and health and social policy. Subsequent conferences have broadened the field to include questions of neuroethics that arise in child psychology, insurance management, journalism, and education, to name just a few other areas.

Today in the United States, health care consumers have access to more medical publications, reporting, advice, advertising, and commentary than ever before. But this abundance has its price: the flood of words and images coming from the airwaves, in print, and in cyberspace threatens to sweep away any pieces of useful information before most health care consumers have a chance to pick them out. This leaves the public not only exhausted but exasperated, and perhaps even more vulnerable than before to the forceful marketing of all kinds of body scans, remedies, and therapies that may or may not be medically sound.

In this context, neuroscientists—who work, after all, on the most complex organ in the body—must make themselves responsible for the public understanding of their research and its importance in everyday matters. This theme emerged clearly at a 2005 symposium on neuroethics, which focused on the most rapidly developing areas in neuroethics today: those having to do with imaging, brain-computer interfaces, and advances in pharmacology. The invitational meeting called together top thinkers in these fields. Co-sponsored by the Library of Congress, the National Institutes of Health, the Columbia University College of Physicians and Surgeons, and the Dana Foundation, the symposium also gave rise to this book. The participants discussed research fresh from the laboratory and its implications for the larger society, addressing along the way many of the questions and concerns about the ethics of brain science that are common among the public today.

Chapter 1 ⌒⌒

A figure that populates many discussions of bioethics in general, and neuroethics in particular, is the chimera—in scientific terms, an organism that carries the genetic instructions for more than one individual. The name "chimera" originally comes from a monster of Greek mythology possessing the head of a lion, the body of a goat, and the tail of a serpent. Is there indeed something monstrous about creating chimeras by implanting human genes into the fetus of a mouse? For an overview of current issues in neuroethics, we hear from two multifaceted thinkers: a lawyer with expertise in genetics and a neuroscientist with expertise in ethics.

WHAT WE CAN LEARN FROM A CHIMERA

Chimeras exist in today's world in a variety of forms. Many of us may be chimeras ourselves, in the sense of having cells of other humans living inside us, says Hank Greely, professor of law and director of the Center for Law and the Biosciences at Stanford University. For example, women who have had children may still retain some of those children's cells that slipped across the placenta into the mother's bloodstream. Some of us, while still in the womb, exchanged a few cells with a twin embryo that disappeared later in the course of gestation. Thus, a genetic mixture of different organisms is not, in itself, new or unnatural.

Even the scientific technique of mixing together genes from different species is not entirely novel; scientists have been making and using chimeras for many years, because for a long time it has been the best (sometimes the only) way to study specialized human cells in a living system. Simply examining them under a microscope or observing them in a petri dish would not have revealed nearly as much about what these cells do or how they do it as we know now. One very successful example of a chimera is a creature called the SCID-hu mouse, which has the immune system of a human being instead of its own murine one. From raising these mice, exposing them to infection, and taking tissue samples from them in order to study their immune responses in detail, medical scientists have gained invaluable insights into the power and versatility of the human immune system.

However, a chimera involving the brain is different. We seem to draw some kind of ethical line at imparting any human cognitive or mental abilities to a nonhuman animal. When Greely and several of his colleagues in an ethics advisory group were once asked their opinion of a research proposal to replace the neural stem cells of a fetal mouse with human neural stem cells, they had no simple answer. In the end, says Greely, the group gave a sort of restricted approval: the researcher was to implant the human genes in the mice during gestation, sacrifice some of them before birth, and closely examine the structures in their brains.

If the brain structures in the mice were clearly mouse structures, the experiment could go forward. If the structures were human-like, or some mixture of mouse and human, the group recommended that the experiment stop for further discussion. For example, because the mouse uses its whiskers as a tool for gathering sensory information about its environment, it has specialized structures in its brain called "whisker barrels" that receive the sensory signals. There is no similar structure in human brains. If the fetal mouse brains in the chimera experiment appeared to be developing normal whisker barrels, the researcher need not intervene; he could allow the mice to be born in the normal fashion. But if the whisker barrels were abnormal or did not exist in these fetal mouse brains, the researcher would need to stop the experiment and return to the advisory group to discuss the implications of this finding. If the whisker barrels were normal and the mice were born, the researcher would have to continue to examine the brain structures and the behavior of the developing mice for signs of possible human influence and would have to stop the experiment for discussion if anything abnormal were observed. The goal of all this monitoring is to avoid the possibility, which the group felt was extremely remote but could not be said to be impossible, that this chimeric mouse might have been given some human cognitive abilities.

However, the likelihood of such chimeras developing to the point of raising any ethical dilemmas is remote, according to Michael Gazzaniga, director of the Dartmouth College Center for Cognitive Neuroscience and a member of the President's Council on Bioethics. Neuroscience is not yet able to identify the neural circuits that underlie any sort of conscious behavior at all. Decades of work have not yet allowed researchers to figure out the neurocircuits that are involved in seeing a triangle, let alone those that account for the higher cognitive functions. Therefore, says Gazzaniga, the concern that a few human neurons injected into a mouse would give rise to a partly human brain has little meaning at this time. "I'm not taken with this concern. The probability that it will work is low, but I also think it's an extreme way of talking," he says, that does not lead to productive discussion of ethical concerns.

ENHANCEMENT, FOR BETTER OR WORSE

One topic that has received a great deal of discussion is the possible ethical distinction between using medication for the treatment of an ailment (therapy) and using it to augment or improve a personal trait or ability (enhancement). The real-life examples of athletes using steroids for enhancement have already provoked enormous debate, but the hypothetical examples offered by neuroscience today are even more contentious; in fact, in some cases, the distinction between therapy and enhancement itself is none too clear. If we raise no ethical objection when a doctor prescribes a drug to treat failing memory in her elderly patients, what do we say when the doctor prescribes the same memory-enhancing drug to a student who is about to take an important exam? Would either of these actions be comparable to prescribing performance-enhancing steroids for an athlete?

Prescribing a pill to improve memory might be akin to giving athletes steroids, according to Greely, since it would be a way to enhance people—not necessarily a bad thing. The ethics of using a memory pill in a specific case would depend on several factors, including whether it was safe and how long it was effective. In the case of a student, for example, if the pill were to work only for the duration of one test and then leave the student once again having to study and memorize material like everyone else, that would be different from a drug that would enhance the student's mental abilities for all the rest of his school years. Another ethical point would center on people's access to the drug: Would everyone in the class be able to obtain it, or only a select few?

Even with these reservations, efforts at enhancement are not always wrong. "I'm a teacher; enhancement is my business," says Greely. "I try to enhance people, both by giving them knowledge and by giving them ways to manipulate knowledge." Gaining knowledge alters the way we think—not in some vague, metaphorical sense but by changing the physical structures in the brain. Learning actually strengthens or creates new synapses, the all-important junctions at which nerve cells pass along their messages to one another, to different sites in the brain, and throughout the body.

Gazzaniga sees the ethical issue of enhancement as having at least two components. In the one that refers to sports, the athletes have implicitly agreed to a deal with their teammates, other teams, sponsors, and spectators. The deal is that the competing teams will all use the same or equivalent sources of food, equipment, and pharmacological aids, or will forgo them if they are not equally accessible to all. If one team uses a resource that the others are doing without, that team has broken the deal. Worse, the team's action contradicts an essential theme of sports, which proclaims the value of persistent hard work.

The second component comes in at the level of the individual: for example, using a drug to enhance memory. The person who does this is making a personal decision to tinker with her own brain, altering her ability to encode information, and does not violate any social contract in doing so. She may or may not realize that such a drug cannot increase her intelligence; it can only increase the rate at which she takes in new information to think about. Unlike steroids, which exist in several different forms and are known to improve athletic performance, drugs that would directly affect the networks involved in intelligent thinking are still only a prospect for the future. For this reason, Gazzaniga considers drugs for athletic performance and drugs for mental processes to represent two different components of the ethical issue of enhancement.

Greely holds, however, that using pharmacology for enhancement is little different from working out in the weight room, having one's diet carefully regulated, seeing a sports psychologist, or going to the best coach one can find. The questions to focus on, he says, are, Should we have rules against pharmacological enhancement, either for sports or for academia—and if so, on what should the rules be based? If the medication poses numerous safety risks, that is a good argument for a rule against it (assuming the rule can be enforced successfully). If there are issues of fairness, with one group receiving the medication and another group not receiving it, that is another reason to think about regulation, either to ban the drug or to make sure that everyone receives it. Many people are wary of enhancement being carried to its extreme, fearing it would undermine our humanity. In that case, says Greely, it would be wrong to carry enhancement to its extreme—but the real difficulty is to find broad agreement on a definition of "the extreme."

Chapter 2

The relatively new technology of neuroimaging can reveal aspects of the brain that have never before been seen, as will be discussed more fully in Part II. Yet, like any other technology, neuroimaging is subject to abuse—perhaps especially so outside the medical context in which it was developed. What are some of the prospects for the nonmedical use of neuroimaging, and how can we tell which ones are appropriate?

NEUROIMAGING AND THE LAW

A fundamental condition of our system of justice is that witnesses must tell the truth. At present, however, the courts have no objective, reliable way to test for veracity. The results from polygraphs, the current lie-detecting machines, are not highly dependable, and many courts refuse even to consider them as evidence. Would some specialized form of brain imaging provide a better way to detect perjury?

According to Hank Greely, several research groups are working on this prospect, so far with limited success. If a truly reliable truth detector can be developed, he expects it will quickly make its way into the courtroom—not by compulsion, but as the courts begin to allow witnesses to take it voluntarily, to establish in a trial of some kind that they are telling the truth. Whether the courts would or should eventually force people to take a brain-imaging test is a harder question that raises a number of constitutional as well as ethical issues.

We might guess that one of the first and most provocative issues to arise would be the potential for invasion of privacy. Here, says Gazzaniga, the nature of brain imaging itself appears to forestall debate.

As will be explained in more detail in the discussion of neuroimaging, a brain scan shows only the pattern of activation in the individual's brain. It conveys a great deal of information about physiological state and signaling activity, but it does not detect and cannot reveal what the individual is thinking. Hence, the possibility of a foolproof lie-detecting system based on brain imaging seems to Gazzaniga to be remote.

A second feature of neuroimaging that stands in the way of courtroom use is that the data vary tremendously from one person to another. Neuroscientists in many laboratories are now studying the extent of individual variation. It has recently been shown that people differ not only in their patterns of brain activation but also in their cognitive strategies for solving specific problems. Brain scans appear to state precisely which neurocircuits take part in each mental process, but in fact what they display are averages (that is, composites made from the scanning results of many individuals) that blur together large individual variations. The purpose of a neuroimaging-based lie detector would be to enable neuroscientists to testify for a defendant's guilt or innocence, according to the levels of activity represented at certain sites in the brain scan. In such a case, though, a neuroscientist for the other side would need only to explain the extent of variation that occurs naturally in a given population, in order to invalidate his colleague's testimony.

Another recent discovery in neuroscience would further undermine the supposed utility of brain scans for unmasking perjury: the brain, it seems, has no specific neurocircuit or signaling pathway for distinguishing truth from untruth. It is reasonable to think, therefore, that a neuroimage would not show a spike in activity levels at any particular site when a person is speaking falsely rather than telling the truth.

NEUROSCREENING AND PREDICTIONS

Many people who value their privacy are concerned that a combination of brain scanning and genetic testing may one day yield predictions about an individual's future health and behavior that could be put to many different uses. The predictions would amount only to statements of probability—they would not have the value of actual information—but this subtle distinction might sometimes go unnoticed. Moreover, it is far from clear who would own or should have access to these not-yet-facts. Should health insurance companies, for instance, be informed that Individual X321 carries the genes for a major disease that is liable to develop in twenty years? Should the local, state, or national police force receive notification that File no. Z789 shows a genetic pattern of predisposition toward violence that might turn to murder at some point?

For some experts, the answers to questions of this kind would depend very much on circumstances. In the example of the late-onset disease, if preventive treatment is available and the health insurance company would pay for it, everyone's best interest might be served by letting the insurance company know. The patient would be spared a serious illness, and the company would have the opportunity to save money by intervening early. However, it would obviously not be in the individual's interest to share such predictions if the health insurance company used them only to deny enrollment or to set higher premiums for people with an identified risk of disease. (One way to resolve the whole dilemma, both Greely and Gazzaniga point out, would be to follow the example of other wealthy nations in establishing a one-payer health system, but that approach falls beyond the scope of this discussion.)

As for the question of alerting the police to someone's genetic predisposition toward violence, Greely says, the answer would depend in large part on the quality of the information on which the prediction is based. Is it based on a study of 80 people that finds a 75 percent risk of some degree of violence, or on a 10,000-person trial that shows that 99.99 percent of the time this genetic signal is followed by significant violent activity? Even with a very solid basis, the proper approach would depend on what kind of intervention is proposed. According to the laws of this country, people cannot be imprisoned simply because they may commit a crime in the future. What about public safety—should neighbors be put on their guard against a long-term risk? Some sort of diversion program, perhaps with an emphasis on anger management or the control of aggression, might be feasible but presumably could not be mandatory. In Greely's view, both the strength of the prediction and the type of intervention would be significant in the determination of who should be notified about what.

Another ethical dilemma related to medical predictions is the question of how a physician should handle the knowledge that a given patient carries a high risk of developing, in ten or fifteen years, a mental or neurological illness for which there is no effective treatment. Is the doctor obliged to tell the patient about this prediction as soon as possible or to protect the patient from this dismal prospect as long as possible?

The best course of action falls somewhere in between, says Greely. He advises the physician first to make sure the patient understands the main costs and benefits involved in hearing a prediction about his future health, and then to let the patient decide whether he wants to hear it or not. As a lawyer in the area of genetics, Greely has seen many examples of the issues raised by scientific knowledge on which medicine is powerless to act. One particularly apt instance is Huntington's disease, a degenerative and ultimately fatal condition of the nervous system. Unlike most genetically encoded ailments, for which people inherit greater or lesser degrees of susceptibility, Huntington's disease is determined completely by one's genetic code; a person who inherits two copies of the form of the gene that causes Huntington's is bound to develop the disease, usually sometime in midlife. Tragically, although medical science has come up with a reliable way to detect the disease-causing gene, there is as yet no known cure, prevention, or treatment.

In Greely's experience, in the fifteen years that the test for Huntington's disease has been available at genetic counseling centers in the United States, only about 30 percent of the people who are at risk and who make contact with a testing center actually decide to come in for the test. This suggests that a majority of us would prefer not to know ahead of time if an unavoidable fatal disease is waiting in our future. And this preference runs deep: even among the people who go to the effort and expense of getting tested for Huntington's disease, about half never return to hear the result. A few prefer to learn their fate and to make their plans accordingly; in any case, the decision to know or not to know rightfully belongs to the individual at risk.

Employers nowadays commonly screen job applicants in a number of ways, from calling the references listed by the applicant to running her driver's-license number through a computer program to check for any criminal records. Will the use of genetic testing and neuroimaging to detect a tendency toward violence or sexual predation or financial misconduct soon become standard procedure in the hiring process?

These intensely personal inquiries would be appropriate only in the narrowest of circumstances, says Greely. If, for example, it becomes possible someday to screen prospective schoolteachers for the risk of pedophilia or police trainees for a proclivity to violence, such measures might be allowable under a strictly defined "health and safety" provision similar to that in the Americans with Disabilities Act. By contrast, candidates for public office should not be required to reveal their medical records—although citizens would be entitled to ask for them and to take the candidates' responses into account when casting their votes. And in fact, as students of American history know well, a good number of our presidents have struggled with significant health problems or other impairments they preferred not to reveal to the public.

Although we may not realize it, predictions play a large role in our everyday lives. Anyone with a young male driver in the family knows that car insurance rates are based on actuarial tables, which are a form of prediction—in this case, a prediction about each driver's likelihood of being involved in an automobile accident. College admissions offices make predictions about an applicant's likely success on the basis of his academic grades, SAT scores, essays, recommendations, legacy status, and perhaps even a secret formula—all of which are of debatable reliability, but which taken together are better than no predictors at all.

By contrast, predictions that are based on genetic background can sometimes be mercilessly reliable, as with Huntington's disease or with Tay-Sachs disease (in which a child born with two mutated copies of the gene will almost certainly not live long enough to enter high school). But more often, genes are only part of the story. Consider identical twins, who carry exactly the same genes but do not always develop the same diseases. As for their behavior, it varies according to their individual experiences. It has to do, says Gazzaniga, with "the fateful chance of different experiences as they develop, given that they have a tremendous amount in common in their bodies and their mental processes."

Therefore, the probability that even a super-neurophysiologist or -neuroscientist can predict a particular person's behavior is fairly low, whereas predictions based on a large number of people are more likely to be accurate overall. In many kinds of predictions, pinpoint accuracy is not required; as Greely points out, a gambler need only be right 52 percent of the time in order to make money.

Chapter 3 ⌒⌒

Some of the most thought-provoking questions in neuroethics today have little to do with day-to-day medical practice. Discussions of what helps or hinders emotional development, the origins of morality, and public concerns about the science of the brain underline the large and growing role for neuroethicists in our society.

TOO MUCH HELP?

Is there ever such a thing as too much therapy? Some observers have expressed concern that the current psychological climate makes life too easy, particularly for young people, by offering prescription drugs that can shield them from ever losing their attention or becoming irritable or sad. Do these kinds of treatment undermine the natural development of the human being to become self-reliant and to overcome obstacles?

For a small number of young people life may be too easy nowadays, Hank Greely says, but for many more it is very difficult. A shockingly high percentage of the children in the United States live in poverty or in troubled homes, or have no home at all. Even the privileged ones who evade these hurdles may need help to overcome disease, an accident, the death of a family member, or some other unavoidable trauma. Appropriate therapy need not interfere with, and indeed can often enrich, an individual's development of character and independence.

Of course, not every negative emotion calls for therapy. Life includes and has always included circumstances in which anxiety is the right response and a lack of anxiety would be dangerous. Michael Gazzaniga describes a study under way in the emergency rooms of the New York City hospital system that focuses on people who have been mugged, asking them whether they have been taking a certain kind of antidepressant; the designers of the study hypothesize that that particular medication subdues anxiety a little too well for some situations. In 1996, psychiatrist Randolph Nesse and ecologist George Williams wrote a book titled *Why We Get Sick,* which argues that the emotional network of the brain has a large repertoire of cues and signals to tell us when to stop overreasoning on the continuum between sadness and depression. These are wired into our brain circuitry because, over thousands of generations, they have helped the human family to survive. Taking a drug that muffles these signals could impair a person's ability to make social judgments, and the possible consequences must be weighed against the emotional, mental, and physical risks of untreated depression. As will be shown below, human decision making relies to a surprising extent on input from the brain's emotional system.

NEUROSCIENCE AND MORALITY

As neuroscientists continue to explore the brain, many people within the field and outside it are wondering whether this work will ultimately reveal something about human nature. For instance, are we all born with an inherent moral sense, or do we acquire such precepts as "you must not kill" only by hearing them over and over again from our earliest childhood?

According to neuroscientist Joshua Greene at Princeton, psychiatrist Mark Hauser at Harvard, and others, virtually all of us tend to react in the same way to certain moral dilemmas (more on this later). This holds true throughout the world, regardless of age, sex, region, culture, or religious belief. Why do we do that? Gazzaniga, whose research focuses on a system in the brain that allows us to interpret our own behavior, speculates that along with emotions such as anger, fear, and jealousy, there may be brain circuits acting directly on our behavior that usually put on the brakes before we carry out an impulse to kill, lie, or steal. We stop, and then because we stop, the interpreter in our brain creates a theory as to why we do that; thus we have a narrative for ourselves and our culture. The abstract notion of morality may have built up from the accumulated wisdom of many such stories until it came to be seen as the main point of these personal narratives rather than an after-the-fact explanation for our behavior.

But another aspect of morality is best illustrated by a simple thought experiment: *Try thinking you are the only person in the world. No other human beings exist. Now, write down all your thoughts of the past forty-eight hours.*

When we attempt the experiment, we quickly notice that the great majority of our thoughts have to do with other people—what they meant by what they said, how we compare with them in one way or another, and the variety of complex social processes that involve us all. We live in constant awareness of the people around us, and perhaps that does constitute a moral compass of some sort.

An exciting line of research in the next twenty years will be figuring out the neurobiology of these kinds of social processes, identifying the brain circuits that enable most of us to coexist peacefully. Just one hundred thousand years ago—that is, five thousand generations—there were only ten thousand people on earth; it is important to keep sight of this point, because it reminds us that we are all quite closely related. We all have more or less the same genes and the same instinctual responses, and neuroscience today is moving toward shining a light on the signaling pathways that produce those responses.

Greely suggests that the place to start looking for universal behavioral norms is not in the brain but in human cultures all over the world. Thousands of anthropologists, sociologists, psychologists, and others have done just that and have found very little in human behavior that looks universal. Even with the traits, attitudes, or actions that are accepted as universal, or nearly so, there remain individuals in every culture who take their own way. Although we tend to label these people as sociopaths, or perhaps as saints, in any case they stand out against any claims of universal human behavior.

But perhaps, says Greely, what is truly universal is the ratio of, say, 95 percent norm-followers to 5 percent standouts. The neurocircuits underlying the norms might be only partially set at birth and still be influenced by early environment and training. If the nature of these circuits is such that 95 percent of children absorb the norms of human behavior and 5 percent do not, maybe it is because that ratio has worked to human advantage over the course of evolution. There could well be some ways in which our species is better off with 95 percent following behavioral norms and 5 percent not.

Among nonscientists, both philosophers and lawyers take a professional interest in questions of personal responsibility. If the brain comes equipped with certain (instinctive or morality-based) brakes on our behavior, what does that mean for questions of legal liability? There have been a few cases in which a defendant was exonerated by pleading, in effect, "my anterior cingulate cortex made me do it"; for the most part, however, neither juries not judges have been very sympathetic to this line of defense. In Greely's view, even if a neuroscientist could prove there was no such thing as free will, we would ignore him in the criminal setting. Most likely we will continue to treat people on trial as if they are responsible, whether we actually believe they are or not, because the principle is so deeply ingrained in our culture.

Drawing a distinction between the abstract notion of free will and that of personal responsibility can make for clearer thinking about each one. Personal responsibility is a social construct. If only one person existed in the world, toward whom would she be responsible? One can be responsible as a social construct with another person—in fact, that is the fundamental rule of social interaction. In Gazzaniga's opinion, a neuroscientist would be mistaken to testify in a court case that a defendant has a reduced personal responsibility because of some particular pattern in his brain scan. The notion of personal responsibility is carried out at the level of the individual, but it does not reside in the individual's brain; it is the rule of the social group. Schizophrenics stop at red lights, he says, meaning that people in all kinds of diminished states, including brain illnesses that cause disordered thinking, can remember and observe social rules. The concept of personal responsibility still has its place in the legal setting.

As for the abstract concept of free will, says Gazzaniga, in a practical sense it means simply that the more educated a person becomes, the more options are available to her brain. Meanwhile, though, all of neuroscience is driving toward a mechanistic understanding of the human brain. With every passing year, we are finding out more about how automatic are the mental processes that we like to consider special to human beings. But even this mechanistic quality does not lessen the social contract that emerges when people come together as a group and agree on social rules.

IMMINENT PROSPECTS AND RESPONSIBILITIES

In 2002, says Gazzaniga, two leading scientists—Daniel Schacter, whose work will be discussed later in this volume, and James McGaugh, of the University of California, Irvine—gave a presentation to the President's Council on Bioethics on a procedure for erasing bad memories. The procedure, which they had tested in animals, involved evoking in the subject's mind the memory of a particular event and then administering a mind-altering drug that eliminated the memory. Some people were fearful of this capability and worried about what it would do to the structure of the individual's personality, but most dismissed it as less than serious neuroscience and expected to hear no more about it.

Just three years later, however, the Cognitive Neuroscience Society held a meeting that attracted more than two thousand professionals and at which more than sixty ongoing or completed studies were presented on this very topic. In addition, a number of research teams were already at work on the development of a pill that could be taken to lessen the emotional power of traumatic memories.

Within a few years, scientists had gone from thinking "This can't possibly be true" to acknowledging that it was true. Now it is time for everyone who worries about bioethical issues to start discussing the specifics of how they want to handle such a drug in society. Difficult issues like this abound in brain science, and they are emerging from the laboratories, the clinics, and the seminar rooms into public debate.

Neuroscientists bear special responsibilities in this debate, says Gazzaniga. They must help the public not only to avoid misperceptions about what neuroscience can do but also to understand what it cannot do. They have a responsibility to place neuroscientific advances and neurotechnologies in context and dispel needless fear about them.

Neuroscientists need to work with reporters and other science communicators to convey the nature and the salient points of their research. In public forums, they should frame and then answer questions such as: What do the data mean? What populations are appropriate to scan? How do we handle resources and clinical ethics demands as neurotechnologies develop and emerge into the public domain, and how do we achieve good education and public engagement?

Because of the expense, ethical demands, logistics, and time constraints, most neuroscientific and neurotechnological studies are one-time efforts. Very few of them are reproducible, and even fewer would be published a second time in a major professional journal. Since this field owes much of its progress to unique experiments with small study populations, neuroscientists need to set a high standard of accountability for their results.

Most importantly, says Gazzaniga, neuroscientists must be responsive to public concerns about the means and the meaning of their work. They need to consider democratic and civic involvement and to promote public understanding of the brain and of brain research. They must be open to everyone's perspectives and at the same time manage the real and imagined uses of neuroimaging, neurotechnology, and psychiatric pharmacology. This volume reflects the idea that the way to reconcile these aims is through widespread inquiry and debate, which drive the progress of neuroethics as well as of neuroscience itself.

"OVERVIEW" (INTRODUCTION AND CHAPTERS 1 THROUGH 3) IS BASED ON A DISCUSSION BY:

Michael Gazzaniga, Director of the Center for Cognitive Neuroscience at Dartmouth College and a member of the President's Council on Bioethics;

Hank Greely, Professor of Law and Director of the Center for Law and the Biosciences at Stanford University; and

William Safire, distinguished author and Chairman of the Dana Foundation.

Part Two

Neuroimaging

Chapter 4 ⌒

Until quite recently, the power to observe a living human brain at work belonged only to mad scientists in horror movies. In the early 1990s, however, the newly developed technology of functional magnetic resonance imaging (fMRI) brought this power into the real world, and the use of fMRI in medical-scientific studies spread at an explosive rate. According to PubMed, the online citation-keeper of record, by 1985 just over two dozen research publications had discussed the prospect of developing a dynamic, real-time form of magnetic resonance imaging; by the end of 2005, the number of research publications mentioning fMRI had grown to more than 146,600. In the meantime, the rainbow images produced by positron emission tomography (PET) caught the public eye and began to make the brain scan a familiar sight in everyday life. Both these techniques reveal the ever-changing patterns of activity throughout the brain. PET does this by means of a low-level radioactive "tag" in the blood (more active areas show greater blood flow), whereas fMRI tracks rising and falling levels of oxygen (more active cells use more oxygen). The technique of fMRI offers less glamour but more anatomical detail than PET and has often supported PET in research.

THE POWER OF AN IMAGE

Despite the numerous, intricate steps involved in the process of fMRI, the end product seems to offer a direct view of the cerebral scene, like the transparent plastic wall of a store-bought anthill. True, the scene from a PET scan is cluttered with information: various areas are delineated, and some specific sites bear labels; colors, ranging from black to blue, green, yellow, and red, represent different amounts of energy consumed, which in turn indicate different levels of brain activity. Nevertheless, the most powerful impression is still the simplest one, that what we are actually seeing is the brain in action. This "picture" is as persuasive as the sonogram of a thumb-sucking fetus, and just as unforgettable.

No one should be surprised that the image of an easily recognized shape, brightly lit up with attractive colors, makes a powerful visual impact. The effect may even be stronger on people who have seen very few medical images before, if any. By contrast, a trained radiologist may scarcely notice the overall picture in focusing on the specific information contained in each pixel.

"When it comes to brain imaging, most people don't understand that they're looking at a mathematical abstraction," says Paul Root Wolpe, director of the Program in Psychiatry and Ethics at the University of Pennsylvania School of Medicine. "Often they also don't understand that PET scans for research are averaged over many individuals, and even clinical scans for specific patients are often put into the frame of someone else's brain." Many quirks of the brain-imaging process are not apparent to the casual viewer. The images seem to be appearing more and more often these days, but with less and less explanation of their contents. Therefore, Wolpe goes on, "I think we have a particular responsibility in the public forum to make sure that people understand what an image is when they see it on television or in popular magazines."

One important point that has been largely overlooked in the wholesale acceptance of PET imaging is that the exact results differ among individuals. The differences are apparent in response to even the simplest stimulus. For the purpose of research, though, the data from groups of study participants are often averaged together to produce a single very clear response. The first decade and a half of PET imaging made great use of the formulas for averaging, while individual differences received relatively little attention. Now, however, imagers can collect higher-resolution data, which will allow them to study individual brain responses in detail. Ideally, this new precision will lead to greater sensitivity on neuroethical issues. "With all the hype as well as the hope," says Ruth Fischbach, director of the Center for Bioethics at the Columbia University College of Physicians and Surgeons, "we must be circumspect about emergent technologies. For many advances, there may be and indeed undoubtedly will be unintended consequences which we must confront."

WHAT ARE WE SEEING?

When brain cells do something, what are they doing, and why is it that we can see a result? One possibility is that the activity represented in a brain scan is all excitation—that is, the release and reception of glutamate at various brain sites may account for all the visual effects that we call activation. In that case, how is its opposite, inhibition—that is, the prevention of cell signaling—visually represented in the same scan? Scientists are currently designing studies that will address such questions.

Something that does not appear in a brain scan, but is important to keep in mind, is the enormous number and variety of processes all surging along at the same time. The brain accounts for only about 2 percent of the weight of the human body, but it uses about 11 percent of the output of the heart and about 20 percent of all of the energy consumed by the body—ten times the usual amount for an organ of that weight. This means that when the data from a brain scan are plotted on a graph, each data point represents a huge amount of brain activity. The irregularities in the curves on the graph, which look like background noise, may hold as much information as the curves themselves.

Individual differences in brain activity do not necessarily produce different outcomes, says Marcus Raichle, co-director of the Division of Radiological Sciences at the Washington University School of Medicine in St. Louis. For any given task assigned to study participants, brain scans may reveal a different strategy being applied by each individual; but, depending on the circumstances and the difficulty of the task, the researchers may or may not observe differences in the result.

Moreover, we cannot take it for granted that activity in the same specific area on several people's brain scans means that these people are all experiencing the same thing. To prove that, scientists would need to carry out what are known as mechanistic studies, in which they probe a very small, defined area, make predictions of exactly what the research subjects should be experiencing, and then check the accuracy of their predictions. They must not reason back from the effect to the cause—that is, they cannot say that because they observe activity in a particular area while a study participant is making a decision, it means that that area is necessarily involved in the decision making. It can mean nothing in itself, since it is only an observation. Of course, scientists must be sure of their observations, but what the observations mean is open to interpretation. In discussing imaging with the public, they must distinguish between direct observations and their interpretations.

The individual differences are very small in relation to the overall metabolic activity of the brain. The moment-to-moment changes are tiny, not like those in a muscle of the arm or the leg when it switches from a resting state to an active state. The brain, when we are sitting quietly, is anything but at rest; neurocircuits in every region are still sending messages by the millisecond to keep us seeing, maintaining our balance, and making sense of our perceptions, all the while monitoring our environment for anything out of the ordinary that might require a new response.

A challenge that remains for researchers is designing ways to study the brain's continuous housekeeping processes, without having it take on any additional tasks related to the study. In other words, how can we understand what the brain is doing when we're not manipulating it?

Chapter 5 ⌒

At first glance, neuroimaging might seem to fulfill the age-old wish of "mind-reading." In truth, brain-imaging technology does not come anywhere near this goal. Nevertheless, it is time for public discussion of the scientific, legal, and ethical issues already raised by the brain-reading technology that exists today.

BRAIN PRIVACY

Not long ago, Jim Holt wrote in the *New York Times:* "The less that is known about how the brain works, the more secure the zone of privacy that surrounds the self, but that zone appears to be shrinking." The issue is whether functional neuroimaging encroaches on this zone, and if so, how much, says Stacey Tovino, visiting assistant professor at the University of Houston Law Center. Some have speculated that health insurance companies, life insurance companies, and prospective employers, to name just a few examples, will all want to use fMRI to "see" how an individual's brain functions in certain situations.

The content of an individual's subjective thought was once an unbreachable barrier, but the momentum of scientific research appears to be wearing that barrier away. Brain-imaging studies can now indicate, for example, whether a study participant has just looked at one category of object versus another. Such advances may possibly point the way to an entirely new form of invasion of privacy.

Does an individual have a right to keep private her tendency to make bad judgments or to conjure up false memories, and would such a right be protected under the Constitution? Tovino observes, "A government-ordered fMRI might implicate the First, Fourth, and Fifth Amendments to the Constitution, depending on the facts involved." In addition, she notes, "The Supreme Court has stated that freedom of thought is the indispensable condition of nearly every other form of freedom, and that the right to think is the beginning of freedom."

The notion of putting neuro-criteria to practical use is likely to appeal to anyone looking for a quick but definitive-seeming way to assess an individual's mental abilities. This prospect raises more than one concern: not only potential loss of privacy, as discussed above, but also the troubling possibility that such criteria may be drawn up by people who have an incomplete or erroneous understanding of the real nature and purpose of brain imaging. A 2005 article by Wolpe and his colleagues in the *American Journal of Bioethics* tells of pressure already weighing on scientists in this field, although their research is still in a very early stage. The studies at this point may involve nothing more complex than monitoring the brains of research volunteers as they deliberately misstate the value of a playing card in their hand; even so, there is pressure from the public agencies funding some of this work to develop the research into a form that they can use.

One prospect that raises ethical alarms is the use of neuro-criteria in hiring practices to screen out applicants with certain disabilities. Some have speculated that employers will use fMRI to look for undesirable characteristics in job applicants, but such practices may violate federal and state employment laws, including Title I of the Americans with Disabilities Act (ADA). Title I regulates the use of qualification standards, employment tests, and other selection criteria that tend to screen out individuals who have a disability, and defines a disability as a physical or mental impairment that substantially limits one or more major life activities. Many conditions that are being studied by fMRI, including depression, schizophrenia, and bipolar disorder, are disabilities if they substantially limit a major life activity, and thus the potential nonmedical use of brain-imaging technology must be weighed with great care.

FALSE MEMORIES

Can brain scans reveal when a person is lying? If so, one possible nonmedical setting for their use that frequently comes up for discussion is the legal system, which depends absolutely on the ability to determine what is true and what is not. (Whether the benefit to society—presumably, the better administration of justice—would outweigh the individual's loss of privacy is also a point of heated debate, of course, as both an ethical and a legal issue.) Daniel Schacter, professor of psychology at Harvard University, points out as an example the well-known problem that inaccurate eyewitness testimony in the courtroom is a common source of wrongful convictions. Another example comes from the long-standing controversy over "recovered" memories, which are notoriously hard to prove or disprove.

Says Schacter, "The whole issue of telling true from false and the implications of neuroimaging for neuroethics obviously is a very big one, and we can think broadly about dividing the domain into two areas. One would be that of deliberate deception, and the question there is whether we can tell through imaging techniques whether someone is engaging in intentional deception or not. The other is the area in which people are doing their best to tell the truth but in fact they are unknowingly lying—that is, they're providing an inaccurate memory report. From the perspective of ethics in neuroimaging, intentional deception and unintentional deception present similar issues."

A certain kind of untrue memory called false recognition can be called forth in a simple procedure, initially reported by James Deese in the 1950s and later rediscovered by Henry Roediger III and Kathleen McDermott at Washington University in St. Louis. First, the participants study a series of semantically related words such as "candy," "sour," "sugar," "bitter," "good," "taste," "tooth," and "nice." Next they take a test in which they are given some words from the list, such as "taste," unrelated words that weren't presented previously, such as "point," and associated theme words or critical words, such as "sweet." The interesting result is that whereas people generally do well at identifying the words that were on the list and stating so with high confidence, and they also do well at rejecting the entirely unrelated words, most of us are quite susceptible to the phenomenon of false recognition—that is, we claim that "sweet," which is related to all these words, was on the list when in fact it was not.

Those related words that were not on the list do nevertheless capture the semantic gist of it, and in a sense this is an accurate memory of what was there—it's just that that particular word was not there. In these cases, Schacter and his colleagues found that a particular brain region, the parahippocampal area, was showing a significantly greater response to the true words than to the false. In fact, the response from this area to the false words did not go above the baseline level of brain activity. Why is that so? Some researchers have suggested the parahippocampal region may be involved in processing context, and other evidence seems to support this. Possibly people are better at remembering the source of the items for the true words, and this may be reflected in the parahippocampal activation.

In a more recent study, Schacter's group tried to ask some of the same questions with a different approach, one they thought would more directly involve the visual cortex. The scientists created a visual analogue of the word-list test for false recognition. They showed people a series of made-up shapes, having ascertained beforehand that people who had seen a number of these shapes were likely to respond "yes" (that is, to have false recognition) to other, similar shapes that they had not in fact seen previously.

The participants carried out the test while lying in a scanner, so that Schacter and his colleagues could study their brain activity at the time they were responding with true recognition of the previously seen shapes and false recognition of the novel shapes. As in the word-list study, the participants were good at saying when an old shape was present but also showed a high rate of false recognition of novel shapes. They showed a significantly higher rate of false recognition of the related shapes than of the unrelated shapes, which again suggests they were remembering something about the general look of the items they had seen previously.

The brain scans revealed that some brain regions have similar levels of activity during true and false recognition, whereas others have different levels. Specifically in the visual cortex of the brain, early regions (thought to be involved in processing primitive visual information from the optic nerve) as well as late visual regions (which handle more complex, partly processed information) are active in both true and false recognition. Both early and later regions show similar responses, which are above the baseline response that appears when people are correctly rejecting items they have not seen before. (These early areas include retinotopic areas and Brodmann areas 17 and 18; the later regions are VA 19 and 37 and the fusiform cortex.)

Differences in brain activity level between true recognition and false recognition appear, somewhat surprisingly, only in the early visual regions. Much as in the word-list study the parahippocampal region shows more activity in response to true recognition than to false, in the shapes study the early visual regions respond more to true than to false. This differential may provide some evidence for a phenomenon called sensory reactivation, in which the early visual regions are repeating the same activity that had occurred earlier when the perception first made its way to the brain. Is this activity associated with conscious memory for the previously studied shapes? Perhaps when a person recognizes one of these shapes and shows increased activity in the early visual region, he is consciously remembering a little squiggle or an odd-looking dent in the shape he had already seen.

The data would suggest that this is not what actually happens. When scientists look at the visual activations for the shapes that really were there, they find that the activation in the early region is the same for those previously studied shapes, both when participants call the shape "old" and when they call it "new." The patterns of activity say nothing about the person's subjective feeling of recognizing or not recognizing something. These data suggest that it is some *nonconscious* form of memory that distinguishes between true and false recognition, and Schacter's research team is now pursuing that possibility.

In other words, after extensive investigation, scientists have not found anything in the brain that looks like a truth-detecting region. Instead, the responses from the many brain regions that they have studied all seem to vary according to the type of test the individual is carrying out, the materials used, the context, and so on. According to Schacter, the application of brain scanning to the real world to prove the truth or falseness of a particular memory simply is not feasible, because of the way the brain is made. "Based on what we and others have done with these techniques in the laboratory," he says, "it appears highly unlikely there is a specific region of the brain that becomes active only and always in the presence of truth."

Chapter 6 ⤳

With its fluorescent colors, evocative shapes, and mind-numbing alphanumeric strings that actually contain crucial medical information, neuroimaging has been serving as the public face of brain research for more than two decades. Even so, 2005 marked a watershed for this field—and for the young field of neuroethics as well. As a young woman in Florida lay in a hospital bed unable to speak or move, brain scans flashed on the airwaves and three branches of government turned their attention to a bitter family dispute and the proper use of a feeding tube. Meanwhile, medical imagers and bioethicists received thousands of desperate inquiries, from all over the country and throughout the world, about the possibility of a brain scan for someone else's critically ill child or spouse, parent, sibling, or lifelong friend. Medical specialists struggled to provide the right balance of hope and caution in response to a tremendous outpouring of public interest.

THERE AND NOT THERE

Amidst the exhaustive press coverage of the Schiavo family's dilemma, a couple of curious concepts took shape that still cloud the public's understanding of neuroethical issues today, says Judy Illes, director of the Program in Neuroethics at the Stanford Center for Biomedical Ethics. One might be called neuro-realism: the notion that when a phenomenon is studied with high technology, such as functional magnetic resonance imaging (fMRI), that phenomenon must be real, regardless of whether the study confirms or refutes it. One recent study ostensibly offers visual proof that acupuncture alleviates pain. In another example of applied neuro-realism, an organization dedicated to the enhancement of moral values has turned to fMRI to prove that pornography is addictive. Neuro-realism can be quickly adopted into neuro-policy, with data from an unreplicated study sometimes being cited as the sole scientific rationale.

The second confusing concept is neuro-essentialism: the sense that subjectivity and personal identity are not just housed in the brain but identical to it. Newspaper headlines or television news flashes such as "Brain Can Banish Unwanted Memories," "How Brain Stores Language," and "The Brain Can't Lie" illustrate neuro-essentialist thinking.

Yet an observation like, "When I make a decision, my brain does the following" is actually a strange thing to say. What does it mean to separate the "I" that makes a decision from the brain making the decision? Clearly, the "I" that is separate from the brain is not just the rest of the body. When one says, "I have a brain," it does not mean that one's kidneys and liver and arms have a brain. It seems we are thinking of some abstracted "I," separate from the brain, as the essence of ourselves. The comment "John is strange" does not mean "John has a strange brain." Instead it says something about the nature or the essence of John.

Nevertheless, says neuroethicist Wolpe, we believe that our brain is who we are. This viewpoint is not universal, however; for example, in Japan, the self is traditionally considered to reside in the thorax. These deeply rooted views hold sway even in our most violent repudiations of the self: whereas a European or an American might carry out a suicide by hanging or by shooting himself in the head, a Japanese person would more likely commit hara-kiri, cutting deeply into the middle of his being.

When we say that someone with advanced Alzheimer's disease

is "no longer there," we are indicating that we think an individual's thoughts, feelings, and interactions with other people reveal the core of her true self. That is why Michael Schiavo could look at his wife Terri Schiavo and see her devoid of consciousness, devoid of will, devoid of desire, sorrow, joy, fear, love, passion—all the things he thought of as vital aspects of Terri—and say that she was gone. At the same time, the Schindlers could look at their daughter and see all the things in her that they loved, feel the closeness of her physical presence, and say, "Terri is still here." Both sides were right: Terri was there and Terri was no longer there. The single most difficult challenge facing neuroethicists today, says Wolpe, may lie in helping practitioners, patients, and their families to sort out this paradox, in which so many emotional, social, philosophical, and emotional issues are entangled.

WHO IS CONSCIOUS?

One of the built-in limitations of the human condition is that, although we know ourselves to be conscious, we can never really know whether anyone else is conscious. A simple distinction based on appearances—the creatures that look the most like us, the human beings, are conscious, whereas the mountains, the clouds, and the caves are not conscious—has long worked for most day-to-day situations. Why does it matter who is conscious? It matters because the moral obligations that we have in this world exist only in relation to other conscious entities, says Martha Farah, director of the Center for Cognitive Neuroscience at the University of Pennsylvania. We have no moral obligations toward mountains or clouds or animal skins.

Nowadays, though, we are confronted with all kinds of more complicated cases in which this simple rule will no longer do. How then should we determine who is conscious? One traditional approach to this problem, originating with philosopher René Descartes, was to reason by analogy. For instance, when I stub my toe I feel pain,

I wince, and I say "Ow." I know I'm feeling pain because I have direct access to that fact. If I see someone else stub his toe and hear him say "Ow," I can reason by analogy and say he is probably feeling pain too.

One problem with this approach is that metaphysically it does not really answer the question. If I assume, on the basis of that other toe-stubber's actions, that he too is feeling pain, or that he is conscious, the question remains, How do I know that?

Another problem with reasoning by analogy is a practical one: it often fails. A clever programmer can design a system that is responsive and communicative and can even say "Ow" under appropriate circumstances but is clearly not conscious. An old-time artificial-intelligence program, "ELIZA," written by Joseph Weizenbaum of MIT, is one example; another is the interactive voice-activated telephone system that tells a customer exactly how much he will need to pay for his FedEx shipment.

Thus, we know that communication is not a guarantee of consciousness. The converse is true as well: just because a system is not communicating doesn't mean it is not conscious. To take one example, a hemorrhage in a part of the brain stem called the pons can cause what is known as a locked-in state. The locked-in patient can no longer move anything except his eyes, although he may still be able to communicate to some extent by means of a computer, owing to recent advances in neurotechnology (which will be described later). Without the computer he would not have that capability, and yet he would still be fully conscious. At this point, reasoning by analogy no longer seems to apply.

With the aid of neuroimaging, however, reasoning by analogy takes on new life. Now, when the hapless subject stubs his toe, in addition to observing him wince and say "Ow," we can observe activity in his anterior cingulate cortex, an area of the brain that processes pain. The analogy is based not only on the subject's external behavior but on the activity levels observed in his brain.

This approach marks an improvement over the simple "ow" analogy because instead of behavior it uses specific and measurable brain activity. Behavior may often indicate a mental state, but it is not the equivalent of a mental state. By contrast, in the field of cognitive neuroscience, there is a real sense in which brain states *are* mental states. From this perspective, the feeling of pain and the activation of the anterior cingulate cortex are two different descriptions of the same thing.

Hence, the brain imaging of severely damaged patients has the potential to tell researchers and clinicians a great deal about the patients' state of consciousness. But there is a catch, warns Farah: before they can make good use of this brain-imaging information, they need to know the relation between brain activity and consciousness. Brain activity that indicates information processing does not necessarily indicate consciousness.

To demonstrate the difference, scientists can look at the results of imaging studies done with normal subjects, in which they experimentally manipulate whether or not research participants are conscious of the stimulus being presented. The researchers can divert the participants' attention over to one edge of a screen, then present a stimulus at the other edge and ask them whether they saw anything. Although the participants will generally say no, nevertheless, if the surreptitiously presented stimulus was a face, it activates the "face" area at the appropriate brain site, and if it was a house, it activates the "place" area. Thus, it is possible to show with manipulations of attention that in normal human brains, unconscious perception goes along with brain activity that distinguishes between different categories of input.

Another kind of study along these lines uses subliminal perception, just as movie theaters used to do with their split-second advertisements for soda or candy. Paul Whelan, of Dartmouth College, has done this, very briefly presenting the image of an angry or fearful face, and has found that even if a person does not know he has "seen" the image, it triggers a characteristic fear reaction in his brain. What this empirical research shows, then, is that finding differences in brain activation that respect differences in meaningfulness of the input does not imply that that input was being *consciously* processed.

"Imaging for consciousness is a new, promising, and ethically important practice," says Farah, "but we need to learn a lot more before it'll be useful. Simply observing that brain activation discriminates among different kinds of inputs does not imply that the brain is conscious."

QUALITIES OF CONSCIOUSNESS

During the months of intensive news coverage of the Schiavo case in early 2005, "My e-mail box was getting approximately fifty messages a day from families of patients who are in the same situation as Terri Schiavo," says Joy Hirsch, who directs the fMRI Research Center at the Columbia University Medical Center. "The messages run, 'My husband [my brother, my cousin] has been in this state; what can we do?' I think it raises the imperative to do the proper research. Sadly, at this point the answer is that we simply don't know how to predict, on the basis of neuroimaging, which patients with severe disorders of consciousness are going to improve and become more conscious."

One cautious estimate puts the number of patients in the United States who are in a persistent vegetative state (PVS) at fifteen thousand, with perhaps one hundred thousand more in a new category, the "minimally conscious state," or MCS. Sobering as these figures are, they cannot be confirmed; in the current incomplete state of medical knowledge, diagnoses of PVS and MCS are based simply on the observation of behavior.

Patients in PVS demonstrate no signs of conscious behavior, although they show spontaneous eye-opening along with evidence of sleep/wake cycles. These patients often do not require mechanical respiration or other life support measures, because the autonomous circuits in the brain stem that sustain these functions have not suffered damage. PVS patients generally do not exhibit sustained, reproducible, purposeful, or voluntary responses to sensory stimulation.

The relatively new diagnosis of MCS recognizes a slightly higher level of behavioral responsiveness than PVS. Patients in MCS can respond, at least intermittently, to simple commands, such as "Move your eyes to the left if your name is Joe," or "Move your eyes to the right if your name is John."

But the observation of behavior falls far short of an understanding of the fundamental mechanisms for loss of consciousness. Fortunately, says Hirsch, investigations of lesions of consciousness or the lack of consciousness are now emerging as a legitimate and productive area of research. One recent study used fMRI in a unique way to assess the neurocircuitry that underlies receptive language in MCS patients. The scientists made a personalized, narrative tape recording for each research subject, with the individual's sibling or parent or other relative coming to the lab to record stories about the circumstances or situations that they had shared with the patient. The narratives, sometimes heartbreaking in their simplicity, ran along the lines of "Remember when you were best man at my wedding," or "Remember when we were children and we used to ride bicycles together." And they brought a surprising result: even in patients who were minimally conscious, the narratives evoked activity in regions of the brain that are thought to house networks for language function in normal, healthy subjects.

One example of this response comes from a patient who sustained a right frontal lobe injury and had been minimally conscious for approximately three years. The patient's brain distinctly shows activity in Broca's area and Wernicke's area—which control the ability to speak and to understand language, respectively—as well as a considerable amount of visual activity, suggesting the possibility that the patient is also picturing the narrative as he hears it. The control group, healthy subjects imaged at exactly the same time, shows very similar activity. With only two patients and ten normal participants as a control group, this study is hardly conclusive, but it raises a new question: If the networks that sustain these cognitive activities are present and active in patients who are minimally conscious, then where is the lesion that has caused their disorder?

Surprisingly, tinkering with the tape recordings themselves helped the scientists to answer this question. Normally the recordings were played forward, of course, but as a control they were also played backward. When the recordings were played normally, the patients and the control group showed approximately the same amount of total brain activity and similar levels in the regions of the brain that would be expected to be active for language. The healthy subjects also recognized the speech that was played backward as speech and even recognized the speaker's voice; they simply could not identify any of the content of the speech, because it was played backward. However, when the MCS patients listened to the recordings played backward, they could recognize neither speech nor voice; they were unable to activate those brain circuits.

Up to now, MCS and related states had been tentatively explained as the product of a cortical network that disease or injury had rendered incomplete. The results of this study suggest instead that at least some disorders of consciousness might be characterized by a very high threshold of activation. They also offer the basis for a testable model of this idea, while raising some compelling questions about the quality of cognition and perception experienced by MCS patients at various stages of emergence.

Although the research described here cannot directly answer such questions, says Hirsch, it holds out the prospect that functional neuroimaging may offer new ways to assess a patient's status and perhaps to predict outcome, as well as lead to a model-based approach that could guide new therapies for the recovery of consciousness.

At present, medical treatment to repair disorders of consciousness remains more of a hope than an imminent prospect. Physicians still have difficulty predicting exactly which patients will emerge and which will not.

Neither researchers nor practitioners wish to give up prematurely the study and treatment of minimally conscious individuals who might have the chance to become more functional in the future. However, Tovino points out, it is important to keep in mind that brain injury rehabilitation is a long, grueling, and costly process for all involved. This includes not only the management of acute injury but also cognitive and physical rehabilitation and/or custodial care. Families and surrogates who contend with the burdens of caring for and worrying about their loved ones suffer from economic stress as well as emotional exhaustion. In addition, many patients who regain additional consciousness will continue to suffer serious cognitive, physical, and emotional disabilities. Some patients who gain additional self-awareness will benefit, but others may endure additional suffering. Patients who are unable to communicate or consistently interact with their environment may be emotionally devastated.

Disorders of consciousness are often considered an orphan topic, according to Hirsch; many interested scientists have been told there is no point in studying patients with these disorders, because they are simply too sick. The research in this field is challenged by an extraordinary number of obstacles, including lack of funding. In many cases, even the routine procedure of consent for these patients is a difficult issue because of a lack of provisions that allow legally authorized representatives to provide it. Moreover, the regulatory guidelines are inconsistent from one institute to another and from one state to the next. Yet this work is never done by one team or one person. These are always massively collaborative efforts, and the inconsistent regulatory guidelines present a major obstacle. In the case of neuroimaging studies, even the simple lack of billing codes for imaging procedures can prove a deterrent.

One message that stands out all too clearly from the tragic case of Terri Schiavo is that our understanding of disorders resulting from severe brain injury falls far short of our imperative to care for the large population of patients who suffer from them. In the future, says Hirsch, advances in the scientific investigation of consciousness and the treatment of disorders of consciousness must form part of the necessary conditions to assure a dignified and personalized treatment of brain-injured patients.

Chapter 7

We all subscribe to the formula that the brain controls behavior, but when we look closely at that formula, it does not explain a great deal. What, after all, constitutes human behavior?

DECISION-MAKING CIRCUITS

In common with all other animals, we have reflexes—but the decisions that guide most of our behavior are clearly different from reflexes. As human beings, we must constantly be able to make new decisions in the face of ever-changing circumstances. From an evolutionary perspective, says Adina Roskies, assistant professor of philosophy at Dartmouth College, the purpose of decision making is to move us to action, so that we behave in ways that will be advantageous to us. Hence, we can't really separate decision making from action. It only makes sense to consider decision making as a highly complex process that goes from appreciating the value of alternatives (at some level, not necessarily explicitly) to choosing one and acting upon it.

Brain circuits for decision making, driven by the prospect of reward, are widespread throughout the animal kingdom. In numerous animals, including ourselves, the brain uses the neurotransmitter dopamine as a reward when certain signaling pathways are activated. The dopamine circuits in the midbrain, striatum, nucleus accumbens, and frontal cortex have been found to be important in this reward system. (This work has been done primarily in rats and in some primates; analogous systems have also been found in insects and many other animals.)

Dopaminergic neurons respond to primary rewards, such as sex and food, but they can learn to respond instead to stimuli that are associated with these rewards. If a particular cue occurs consistently with a reward, the dopaminergic neuron will initially respond to the reward but will eventually respond to the cue instead. The neuron will also show a depressed level of activity if the reward that it expects fails to occur. This type of responsiveness to the contingency of cue and reward is fundamental to the reward system that mediates learning.

In addition to the midbrain dopaminergic system, the ventral medial prefrontal cortex (VM cortex) is involved in associating complex stimuli that are not primary reinforcers with reward signals. This area of the brain plays a key role in figuring out which signals are significant for possible reward when the rewards themselves are unpredictable—that is, when circumstances change.

We humans like to think of ourselves as the most cognitive of the creatures, the most able to think about abstract topics. (The humorous philosopher Mark Twain made this point from a slightly different perspective: "Man is the only creature that blushes—or needs to.") But the faculties that we use for making decisions are not uniquely human at all, Roskies says. They are the abilities to calculate the chances that a given contingency will occur, and to estimate the value or reward of each contingency. As mentioned above, a great many animals have these abilities as well.

Most recently, neuroimaging studies have shown that not only computation but emotional processes are important in good decision making. With so many brain circuits apparently involved, which parts of the brain are doing what when we say we are "making up our minds"? At the University of Iowa, Antonio Damasio and his colleagues developed the Iowa Gambling Task to study how people make decisions under conditions of uncertainty or incomplete information.

The participants were given four decks of cards from which to pick. Unbeknownst to them, two of the decks had very large gains and occasional large losses, whereas the other two had smaller gains and smaller losses. The value of the gains and losses was manipulated toward a particular outcome: over time, choosing from the two decks with the smaller gains and smaller losses would win the player more money than choosing from the larger-gain, larger-loss decks.

The study found that normal humans tend to perform well on this task. Although they know nothing at the outset about which decks give better or worse results, shortly after they start picking cards they come to understand that the two decks that give smaller rewards are the better decks to choose from. In addition, they develop a subtle but measurable physiological response, known as a skin conductance response, toward the bad decks. It is when this skin conductance response emerges that they start choosing from the better decks. If the players are asked at that point whether they understand which decks are good and which are bad, they say no; they do not yet have explicit knowledge of the pattern. Nevertheless, the skin conductance response is already evident.

In contrast, participants in this task who have sustained damage to the ventral prefrontal cortex or orbital frontal cortex of the brain do not develop a skin conductance response. As described in a paper by A. Bechara and others at the University of Iowa, they also fail to choose correctly; that is, they continually choose cards from the worst decks. After playing longer, some of the patients can state explicitly which decks are good and which are bad—yet they keep choosing from the decks they know are bad. Their cognitive understanding is dissociated somehow from their action; cognition alone does not enable patients to improve their performance. Interestingly, most of us would say at this point that these patients are making poor *decisions* rather than taking *action* poorly—an illustration of how, at some level, we recognize that making a decision and taking action are two points in the same process.

These results underscore the importance of the ventral medial prefrontal cortex in good decision making, says Roskies. Further neuroimaging studies along these lines in normal human subjects have found that the ventral medial prefrontal cortex and the orbital frontal cortex are regularly active in tasks of this kind. These brain areas are important not only for associating outcomes with reward but also for integrating sensory and deliberative information with signals from the limbic system, which mediates emotion.

Moreover, this work has revealed activation in brain areas that are part of the reward system as well as in the amygdala, a brain site that is responsible for the perception of threat and risk. (Some researchers have therefore suggested the amygdala is involved in predicting bad outcomes in rather a general sense.) Neuroimaging from these studies has also shown activity in the insula, a part of the cerebral cortex known to switch on in situations that carry a high risk of punishment, of which monetary loss is but one example. The dorsal lateral prefrontal cortex manages the online manipulation and integration of decision-relevant information. One more area found to be active in this context is the anterior cingulate cortex, which is involved with conflict monitoring and risk. The interchange of signals among the sites named above, together with the brain's networks for working memory and for deliberative processes, all figure into the neural circuitry of decision making.

MORAL DECISION MAKING IN THE HUMAN BRAIN

The research just described has yielded many insights into the brain mechanisms that underlie decision making, but most have come within the confines of decisions about the gaining or losing of money. Can this really represent the summit of human reasoning? If there is one form of decision making that we think is entirely abstract and therefore uniquely human, it is moral decision making; how does the brain handle that?

In the realm of research on moral decision making, people who have suffered brain damage in the area of the VM cortex are celebrities. They are famous for being able to identify the right thing to do in a given situation but not doing it. From a philosophical standpoint they are interesting, says Roskies, because one well-known philosophical theory holds that a person who knows the right thing to do is thereby motivated to do it. VM patients appear to offer a counterexample to that theory.

When, as a thought experiment, both VM patients and normal subjects are asked to consider certain moral dilemmas, an interesting pattern emerges. One classic dilemma is the trolley problem: suppose you see a trolley headed for five people who are standing on the track. If you do nothing, those five people will die. However, there is a side track on which only one person is standing, and if you flip a nearby switch you can divert the trolley from hitting the five to hitting just the one. Should you flip the switch? Most people say the right thing to do in that case is to save five lives at the expense of the one. In this instance, flipping the switch is the moral choice.

But now suppose the trolley is headed for the five people and there is one man standing on a footbridge that runs over the track. He happens to be a very large man, so that if you push him off the bridge and onto the track, his bulk will stop the trolley before it hits and kills the other five people. Should you do it? Most people say absolutely not. Then, says Roskies, we must ask, Why is it morally acceptable in the first case, but not in the second one, to kill one person for the sake of saving five others?

Consider the same two scenarios, this time in a medical context: you are a surgeon. You have five patients who desperately need surgery today, and if you operate quickly on each of them you can save all five. Your sixth patient, however, needs a life-saving operation that would take the entire day to perform. As in the trolley problem, most people vote to save five people at the expense of the sixth. Now, to give the scenario another twist, suppose you have five patients who urgently need organ transplants in order to survive, while a sixth patient has the healthy organs that the other five patients need but urgently needs a different organ himself—and suppose that sixth patient has agreed to be an organ donor. Should you sacrifice the one to give his organs to the five? Most people say no.

What is taking place in the brain while we ponder these purely hypothetical questions? Joshua Greene, of Princeton University, and his colleagues have carried out numerous imaging studies of the brain wrestling with such dilemmas, and they have noticed something startling: the regions that are active in these moral judgments are the same regions that are active in general decision making, and even in decision making about monetary risks. According to data from neuroimaging, the brain has no moral center. The kinds of sites and circuits that we use in order to make judgments in general are the same ones we use to make moral judgments. Slight differences may exist, but this study found no signs of a specialized brain area for moral decisions.

What is abundantly evident from Greene's neuroimages, however, is that when we consider such difficult questions as whether to push the large man off the bridge or whether to let one patient die in order to distribute his organs to five others, the emotional areas in the brain take on a greater role. Something is motivating us, not logically but emotionally, when we say, "No, you can't do that, because it would be wrong."

What is the right way to answer a hypothetical moral question? Neuroimaging cannot tell us that, says Roskies. It cannot tell us what we should do. It can only tell us what we do and how we do it. Clearly, neuroimaging has provided and continues to provide insight into how humans make decisions by using brain areas that are shared with other animals as well as areas that are uniquely human. It has revealed to us that decision making is more than just a cognitive affair, that it involves emotion. This insight in particular may help us to begin to understand how human morality evolved from more simple systems.

"NEUROIMAGING" (CHAPTERS 4 THROUGH 7) IS BASED ON PRESENTATIONS BY:

Martha Farah, Professor of Psychology and Director of the Center for Cognitive Neuroscience at the University of Pennsylvania;

Ruth Fischbach, Director of the Center for Bioethics at the Columbia University College of Physicians and Surgeons;

Joy Hirsch, Professor of Psychology and Functional Neuroradiology and Director of the fMRI Research Center at the Columbia University Medical Center;

Judy Illes, Director of the Program in Neuroethics at the Stanford Center for Biomedical Ethics, Stanford University School of Medicine;

Marcus Raichle, Professor and Co-director of the Division of Radiological Sciences at the Washington University School of Medicine in St. Louis;

Adina Roskies, Assistant Professor of Philosophy at Dartmouth College;

Daniel Schacter, Professor of Psychology at Harvard University;

Stacey Tovino, Visiting Assistant Professor at the Health Law and Policy Institute, University of Houston Law Center; and

Paul Root Wolpe, Professor of Psychiatry, Medical Ethics, and Sociology and Director of the Program in Psychiatry and Ethics at the University of Pennsylvania School of Medicine.

Part Three

Drugs in the Brain

Chapter 8

In view of the millions of prescriptions that are written every year for drugs that act on the brain, it is startling to remember that there are no objective diagnostic tests for mental disorders. Currently, no blood test can tell an individual whether she is at risk for major depression or an anxiety disorder or whether his attentional problems are past a certain threshold. The *Diagnostic and Statistical Manual of Mental Disorders (DSM)*, the diagnostic manual for psychiatry, suggests that various collections of symptoms each add up to a distinct psychiatric disorder, but the reality is that depression, attention deficit disorder, autism, and probably many other conditions are more like quantitative traits, which can exist to a greater or lesser degree. An individual may exhibit several symptoms, some of them very clear and others less so. With each case, psychiatrists are called upon to draw a line somewhere in the continuum of symptoms, distinguishing people who are in need of treatment from those who are not.

Sometimes, with advances in medical science and technology, the borders themselves are subject to change. This can happen when the risk of a certain disease is found to be significant in the presence of milder symptoms or signs than had previously been highlighted. An example, says Steven Hyman, provost of Harvard University and professor of neurobiology at Harvard Medical School, is the circulating blood levels of LDL cholesterol. Epidemiology and clinical trials of cholesterol-lowering agents have led to substantial changes in the threshold for treatment. A change in the borderline for treatment can also come about when new research identifies a previously unnoticed red flag, as when it became clear a few years ago that not only high diastolic but high systolic blood pressure, on its own, constitutes a risk factor for stroke and might require treatment. Borders in the continuum between health and illness in the brain are perhaps particularly subjective, but in fact most of medicine is typified by diagnostic gray zones. Very few conditions exist in which one can simply sort patients into the sick and the nonsick.

Physicians today work in a world in which everyone, from specialist to generalist to layman, seems to have an opinion on many medical topics. One widespread opinion, unfortunately, is that interventions aimed at addressing mental disorders differ in some profound way from interventions that aim to address general medical disorders. This opinion is especially strong when the intervention ought to take place as soon as possible, but the risk hovers over the distant future. To take a hypothetical example, a teenager with a dangerously high level of cholesterol (say, around 300 mg/dL) would most likely receive a prescription for a statin immediately and for many years, even in the absence of any data on the long-term effects of statins on childhood development or in the decades afterward. However, the psychiatric equivalent—such as antidepressant drugs for a young teenager who shows two or three depressive symptoms and therefore carries a significant risk of clinical depression in the next few years—would be met with suspicion, at best. If in both cases the youngster is currently fairly healthy and the consequences of long-term drug treatment are unclear, why are the cases regarded so differently? The answer may lie partly in the categorical difference we persist in seeing between the brain and all the rest of our organs—a largely Western point of view, as mentioned in Part I.

STARTING WITH SAFETY

The public-health community, whose primary goal is the prevention of large-scale disease, has endorsed the use of statins to combat heart disease. Still, some of the traditionalists worry about another kind of risk, a moral hazard: Will a number of the people who take statins shed their willpower and come to feel they can eschew exercise or eat anything they like without regard to its effects on their health?

Most of us might ignore the theoretical risk or else accept it for the sake of lowering our risk of heart disease. But this abstract point cannot be so easily waved away. Unexamined, it reappears in other aspects of our thinking about the proper role of medicine in our society—and might even bear some weight in the shaping of medical policy. The notion is that effortful engagement, whether psychotherapy, physical training, or other kind of arduous workout, carries a moral benefit, whereas taking a pill is morally slovenly and even harmful—a moral shortcut. (Gerald Klerman, longtime professor at Harvard Medical School, referred to this belief as "pharmacological Calvinism.") The brain itself makes no distinction between pharmacology and experience. In this teeming, continually changing network, every physical, mental, or spiritual exertion leaves its mark in the fine-tuning of synapses and neurotransmitter levels, as does every pharmacological intervention.

One reason that the use of statins as a preventive measure has become very common in this country is that we assume these drugs are both safe and very effective. In fact, numerous studies have shown exactly that, and it was only after they had done so that a consensus on statins began to form among the medical community and the public. If they had not been proven to be both effective and safe, "statins" might not now be a household word.

The necessity of demonstrating these two qualities may seem almost too obvious to mention; but Hyman points out that, as we should remember from recent experience with one type of antidepressant that carries a small but real risk of enhancing suicide-like behavior, we cannot take either quality for granted. Drugs that have become familiar may not be as safe as we used to think; then again, we should not alarm ourselves into withholding them from people who are really ill.

One problem in drug development often overlooked or unknown among consumers is that we have relatively little data so far from clinical trials, especially for children. In the trials of drugs aimed at treating a disease, most studies run for six months or less—even if the drug under study is intended for chronic treatment that may continue for years. Hence, a drug deemed safe is known to be so only over a certain period of time, explains Russell Katz, director of the Division of Neuropharmacological Drug Products at the U.S. Food and Drug Administration (FDA). After this period the drug may remain safe to use indefinitely, or may gradually come to carry more risk (of, say, deleterious side effects), or may even, under certain circumstances, become unsafe with long-term use. In the first stages of clinical trials, the term "efficacy," too, applies only within certain limits: the drug need not produce the intended effect 100 percent of the time, or even 50 percent of the time. Especially in the case of life-altering disorders such as epilepsy, depression, or Parkinson's disease, a drug need only work better than a placebo—that is, better than no treatment at all. This point may not be emphasized in a drug advertisement, but it can usually be found somewhere in the small print of the ad or in a packaging insert, which is formally considered to be part of the label.

When drugs are used for people who have milder symptoms or are at lower risk, or to enhance already normal abilities, standards of safety have not yet been developed, says Hyman. To date, the medical and pharmaceutical communities have not formed a consensus on the kinds of studies that would be required to develop standards for prevention in diseases that are not life-threatening. How large would the studies have to be, and how long would they have to run, in order to prove a drug safe and effective in people who are healthy to begin with? Would they have to be placebo-controlled and run for five years? Would they have to have ten thousand patients? We don't yet know the answers to these questions; indeed, we have only recently begun to discuss them.

What features should the FDA take into account when weighing whether or not to approve a drug that might be taken by healthy people to augment or improve some aspect of themselves? Again, safety is the most important consideration, says Hyman. A doctor about to prescribe a drug for someone who is not sick must be confident that the medication itself will not make the person sick.

The proposed labeling of a drug on its way to market receives as much scrutiny as the drug itself. In fact, according to Katz, drug approval is inextricably linked to labeling. The law requires "substantial evidence"—usually at least two well-controlled investigations, including clinical studies—demonstrating that the drug produces the effects that are claimed for it on the label. From the standpoint of legal requirements, developing treatments to enhance normal function is permissible as long as the research is solid and the labeling describes exactly what the drug does and for whom it is intended.

Would we be willing to relax the safety standards for an enhancement drug that produces a very substantial effect, catapulting the user from, say, average intelligence to brilliance in one dose? We don't yet know the answer to this question, since no such drug exists. As of this writing, says Katz, the FDA has not received notice of any pharmaceutical company working along these lines. True, we already tolerate some treatments that cause considerable harm—we have only to think of amputation, radiation, or chemotherapy—but these measures are reserved for patients who are gravely ill, and they are known to convey far more benefit than harm. Whether healthy people will risk endangering their health for the sake of mental improvement remains to be seen.

PSYCHIATRIC DRUGS FOR CHILDREN

Entering children into clinical trials of a new drug for mental disorders raises its own set of ethical questions. For one thing, all the data that scientists have up to now have come from trials of drug treatment for recognizable disorders. Little information has been gathered on the early intervention for or prevention of, say, the onset of depression or an anxiety disorder, and none at all on the enhancement of capacities outside of very small laboratory-based studies that don't necessarily mirror real life. Therefore, when we give drugs to a child, especially one who may not be ill, we ought to be clear about whose interest is being served by the treatment. Is it in the interest of society, the child's school or parents, the child himself, or some combination? Hyman points out that these interests are often difficult to distinguish from one another.

Then there are children who have an accepted medical diagnosis, but are they diseased? Pediatric psychiatrists face this question frequently in their practice in a syndrome known as pediatric conduct disorder. This diagnosis, the most common reason for referral to a pediatric psychiatrist, has as its essential feature a persistent pattern of behavior in which the basic rights of others or major age-appropriate societal norms or rules are violated. In other words, this condition is diagnosed not by the patient's own health or well-being but by the effect that the patient's behavior has on other people. The criteria include using a weapon, setting fires, running away from home, and so on—all kinds of behavior that trouble other people for the most part, but not necessarily the patient. Another criterion is that this behavior interferes with the patient's functioning—but of course, if someone sets fire to a house, stands trial, and is sent to jail, that will indeed interfere with her functioning. Clearly, whether or not it is caused by a disease, such conduct is something that needs to be addressed, says Thomas Murray, president of the Hastings Center. Psychiatrists as well as lawyers and judges have a responsibility to think about the larger implications not only for that person but for all the others on whom that person's life will touch.

The criterion of "interfering with the patient's normal functioning" merits a closer look. One might reasonably argue that the way our society responds to the troubled and troublemaking child, whether by isolating him or by rushing to medicate him, interferes with his normal functioning almost as much as a latent psychiatric disorder could do. This may be especially true in cases in which the child shows a few warning signs but not a full-fledged disorder.

Pediatric conduct disorder, like a number of other pediatric psychiatric problems, presents special ethical issues because the subject is a child, whose brain is still developing and therefore more subject to influences from the environment than the brain of a mature adult. Drug treatment, behavior modification, or whatever else is used to manage the disorder in a fifteen-year-old is likely to affect her future profoundly.

Many children who do not receive a formal diagnosis are nevertheless struggling with psychiatric symptoms: for example, the eighth-grader who has lost his appetite, hardly sleeps, and feels guilty all the time, or the sixth-grader who is very bright but does poorly in school because her restlessness keeps her from listening to the teacher. All too often, says Hyman, these children drag themselves through years of underperformance in school, think little of themselves, and are rejected by most of their peers. Perhaps they feel, too, that their teachers or parents see them primarily as problems. They are likely to turn to other self-styled outcasts, from whom they may learn such maladaptive habits as avoiding school entirely or self-medicating with alcohol or illicit drugs. The luckiest among them will be caught, undergo medical examination, and finally receive a diagnosis for what has been hindering them all along. The saddest part of these stories is that even the best psychiatric help in the world cannot make up for their lost years of scholastic and social development; they must start anew from where they are now. Just as surely as treatment with powerful drugs, experiences that result from a lack of treatment etch their marks in the brain. This fact alone may offer enough reason to promote early intervention to reduce the impact of psychiatric disorders in children.

UNFAIR ADVANTAGE IN A PILL?

Some time ago, when the names "Ritalin" and "Prozac" still sounded strange to most Americans, a few people warned that these drugs, which altered the state of the brain, could one day be used to control large numbers of people for political ends. Fortunately, this potential has not been realized; instead, the opposite problem has developed. According to several studies and surveys, Ritalin and Prozac are used at highest rates among advantaged Caucasian populations. The issue now, according to Hyman, is, Are we allowing the emergence of two classes, the chemical "haves" and the "have-nots"? Are we going to create a wealthy, comfortable class whose children will have not only the latest computers and special college-preparation courses but pharmacological advantages for competing in school?

This is not a trivial ethical issue for the kind of society we want to have. We need urgently to discuss such questions and come to an understanding. The drugs now in development have the stated goal of treating illness, but almost inevitably some will prove effective for enhancement and will be used by healthy people who can afford to pay for them. In our free society, many believe it is their right to do whatever they can to minimize their distress and maximize their achievement. They may believe it is their duty to give their children every advantage. The medical establishment cannot effectively prohibit these uses or persuade all physicians to turn away such requests. We need, therefore, to address the ethical issues at stake so that we can set some guidelines to manage the ever-growing demand.

There is no question that the use of psychotropic medication has increased dramatically over the last ten to fifteen years. At the top of the list have come the stimulants, particularly methylphenidate (Ritalin), and amphetamines in general, which are used to treat the condition of attention-deficit/hyperactivity disorder (ADHD). ADHD is characterized by a level of inattention and/or hyperactivity that is abnormal for the child's stage of development, that usually becomes apparent in preschool or early elementary school years, and that impairs school performance, can lead to academic failure, raises the risk of delinquency and of accidents, and increases overall medical costs. This list of consequences does not mean each child diagnosed with ADHD is doomed to have these negative outcomes, but on average, a population of children with ADHD would have more of them than a comparable group without ADHD. The prevalence of ADHD in American children is estimated at 4 to 5 percent and is more common in boys than girls.

Medical treatment for ADHD frequently includes the use of stimulants, but exactly how this class of drugs works on ADHD has never been settled. Benedetto Vitiello, chief of the Child and Adolescent Treatment and Preventive Interventions Research Branch of the National Institute of Mental Health (NIMH), says it is unclear whether stimulants have a beneficial effect on educational achievement in addition to enhancing cognitive performance. There is plenty of evidence that these drugs increase alertness, attention, and goal-directed activity while reducing fatigue, appetite, and sleep. The evidence is less clear on the question of whether they increase knowledge, accelerate learning, or bring about better academic achievement in general. Experts agree that stimulants improve performance on tests; that is, whatever a person is doing, he or she will do better under the influence of stimulants.

According to a 2002 estimate by Ann Zuvekas and her colleagues at the George Washington University School of Public Health and Health Services, between 2 and 2.5 million children in the United States received a prescription for stimulant medication at least once that year. Large as this figure is, it may be an underestimate, because it is based on reports by parents. Although the study population was carefully selected to represent the nation as a whole, the results of these surveys can be subject to over- or under-reporting.

One recent survey attempted to identify the main factors that might raise a child's likelihood of using stimulants. As might be expected, age is a powerful predictor: children ages ten to twelve show the highest rate of use. Other factors include being a boy, living in a family with fewer children, and living in a community with a higher percentage of Caucasians. Rates of stimulant use among children vary widely with geography: higher in the South and Midwest, lower in the Northeast. Among the states, the lowest rate of use is found in Washington, D.C. (1 percent), and the highest in Louisiana (about 6 percent).

An interesting study, from Adrian Angold and his colleagues at Duke University, concerns a survey taken in the late 1990s in western North Carolina. Not content with simply counting how many subjects were taking stimulants, they wanted to find out to what extent the children's symptoms, as reported by parents or teachers, actually were consistent with a diagnosis of ADHD.

The Duke scientists studied an epidemiological sample of about fifteen hundred children, ages nine to sixteen, for the period from 1992 to 1996, and found that about 3.4 percent of this number—about fifty children—met full criteria for ADHD (a figure that is in accord with most such surveys). Of these fifty or so, almost three-quarters had taken medication consistent with their diagnosis—that is, stimulants. Another group, representing about 2.7 percent of the study population, had some symptoms of ADHD but did not meet the full criteria for diagnosis; about a fourth of them used stimulants. The great majority of the study population, more than 93 percent, showed no symptoms suggestive of ADHD at all—yet, within this group, about 4.5 percent had used stimulants at least once in those four years.

Focusing in on this 4.5 percent yields a few additional insights, says Vitiello. It appears that, of a given one hundred children in the community who were treated with stimulants in the four years, about 34 percent had symptoms that met the full criteria for ADHD; about 9 percent had symptoms but not full-fledged ADHD; and about 57 percent had neither. However, about 84 percent of the children who had no ADHD symptoms were nevertheless rated by their teachers as impaired in school.

Another epidemiological study, called Monitoring the Future, was sponsored by the National Institute on Drug Abuse. This was a survey carried out on a representative sample of eighth-, tenth-, and twelfth-grade students. The response (again, by self-report) indicated that some 4 percent of those queried had taken Ritalin without a doctor's order at least once in the preceding twelve months. Several factors stood out as predictive: being white, being in a higher grade, having low school grades, and having a history of substance abuse. Some geographical variation in these rates may indicate this type of use was carried out not to improve school performance but for recreation.

In Vitiello's view, stimulants have a wide range of possible applications, from use to misuse to abuse. The favorable ratio of benefit to harm that has often been cited for stimulants has been found specifically for children who meet the full criteria for ADHD; no one knows just what the ratio is in other groups of stimulant users, such as children who have symptoms of ADHD without the full diagnosis. Yet, treatment without a full diagnosis is not unheard-of in medical practice. In fact, most medications are used at times in subsyndromal cases or in cases with a different, though related, diagnosis—not just in psychiatry but, for instance, in cases of antibiotics being prescribed for infections that may not be caused by bacteria.

Moving to the next point on the range of applications, stimulants may be used for treating children who do not have symptoms of ADHD but may have other learning disabilities such as dyslexia or even a level of aggression that keeps them from focusing on their schoolwork. There is some evidence that treatment with stimulants improves these children's performance; whether they actually learn more is still an open question. Another example of nontherapeutic use is the practice of giving stimulants to a healthy, nondisabled child to improve her performance in school.

The endpoint on the range of applications, taking stimulants to bring on euphoria, is clearly drug abuse. The two nontherapeutic uses that fall short of that may arguably constitute misuse; if so, it is a misuse being carried out millions of times every year. The arguments for and against the nontherapeutic use of stimulants each have their passionate advocates. However, the key element of this debate—solid evidence about the risks and benefits—is missing, because the necessary research has not been carried out in healthy children. Until this has been accomplished, no argument on either side can be entirely convincing.

Chapter 9 ∽

In deciding how to treat a neurological or psychiatric condition, physicians must weigh the possible benefits of each option against its possible harm; sometimes the solution is difficult to see. Understanding how a prospective drug treatment is likely to interact with a patient's genetic background can help to make the answer clear. Philosophical issues of whether the drug offers treatment or enhancement, justifiably or not, complicate individual cases but contribute to the larger social discussion.

WHAT IS WORTH TREATING?

Overshadowed by the heated debate about the use of therapeutic drugs in the absence of disease, there simmers another question: Does every disease merit treatment? For that matter, what constitutes a disease? This may seem a simple-minded question, but Katz offers three instances in which the answer is far from obvious.

The first instance involves health conditions that are probably normal for older adults but are not normal for younger people. One such condition is age-associated memory impairment, or AAMI, in which elderly individuals don't have the excellent memory of people born thirty or forty years later, but their memory is normal for their age. Depending on the definition of AAMI, this condition may affect 70, 80, or even 90 percent of the population above a certain age.

At such high prevalence, AAMI might well be considered normal. Is treating AAMI then considered therapy, or is it enhancement? No one can say for sure, but what is certain is that drug companies are actively developing treatments for it.

The second instance includes ailments that might be described as episodic rather than continuous. One example is migraine. In people who are subject to these very severe headaches, often with dizziness, nausea, and other disabling symptoms, they tend to occur a few times per month and to last for several hours or even a day, and occasionally longer. A prophylactic drug would have to be taken every day throughout the year. What would be the proper ratio of risk (of side effects or damage from long-term use) to benefit (preventing most migraines)? Seasonal affective disorder is another example: how safe does a drug have to be for wide use against clinical depression for several months of the year if, as the scientific evidence suggests, only about 30 percent of people who are susceptible would actually suffer depression if left untreated?

The third instance has to do with currently healthy people who carry a known risk of some particular disease: first-order relatives of patients with schizophrenia, depression, or early-onset Alzheimer's disease, for example. If a preventive drug existed, how many normal people would have to be treated in order to bring the prevalence of the disease down to an acceptable level? Since medical science has not yet developed strong predictions of who, among all the people at known risk for a disease, will actually develop it, how long and how large would the clinical trials have to be to allow preventive drugs on the market? A medical consensus on these matters, or even on an objective process by which to examine them, would do much to help set priorities for research funding.

HOW GENES INTERACT WITH DRUGS

An important element for discussing ethics in neuroscience is an understanding of how an individual's genetic makeup contributes to the effects of any kind of drug he may take. Without having this notion in place, we risk missing out on a particularly lively corner of the research field where neuroscience, genetics, chemistry, and pharmacology form a productive partnership.

It is well known that the human genome, or entire set of genes, contains innumerable differences from one person to another. And of course genes have a lot to say about each person's innate abilities and vulnerabilities (although they are rarely the sole determining factor). Hence, says Daniel Weinberger, director of the Genes, Cognition, and Psychosis Program at the NIMH, drugs or other measures to enhance cognition will meet with a different response in each individual, based on his or her unique genome.

Up to now, most efforts to enhance human cognition by means of drugs have been judged by their outcome. The idea has been that the individual takes the drug and somehow, in the unfathomed recesses of the human brain, the drug miraculously changes the person's performance on a test of cognition. Not long from now, however, the study of human genes will identify the basic mechanisms by which these cognitive abilities develop and are modulated. This knowledge will lead to more precise strategies for changing cognition, which will vary with each person's genetic predispositions and susceptibility to side effects.

To illustrate how drugs interact with genes in the complex process of human cognition, Weinberger offers the example of a specific gene he and others have studied. This gene makes a protein that plays an important role in working memory, problem solving, and abstract reasoning. It is known as COMT because it encodes the instructions for making the protein catechol-O-methyltransferase, which in turn helps to regulate the amount of dopamine that serves as a signal between cells in the executive cognitive processing region of the brain—that is, the region in charge of planning, organizing, decision making, and attentional activity.

In this region, the prefrontal cortex, the neurotransmitter dopamine has the job of fine-tuning the way the brain processes information. A certain level of dopamine was found by Patricia Goldman-Rakic and others to be optimal for this task. Too low a level of dopamine is associated with the development of Parkinson's disease; too high a level presents the risk of schizophrenia. According to Jeremy Seamans, of the Medical University of South Carolina, the non-optimal dopamine state in the cortex is one in which many environmental stimuli jostle for access to our conscious attention. The optimal level of dopamine allows the brain to identify critical information and put it to use in a conscious, ongoing representation of the environment. It is this representation that guides human behavior and allows us not to be distracted by other information that is not relevant or not helpful at the moment. Dopamine enables a neurocircuit to maintain its signal, or its attention to a target.

Equally important for day-to-day functioning, of course, is the neurocircuits' ability to shift to a new target when necessary, and this is the role of COMT, which inactivates the dopamine synapses. (The identification of the COMT protein marked a great advance in neuroscience and earned its discoverer, Julius Axelrod, of the NIMH, a Nobel Prize.) The COMT gene exists in two forms, which differ only by a single change in the DNA sequence. This change appears to have emerged about five million years ago in a hominid—that is, prehuman—ancestor, whose sperm or egg simply misspelled a certain DNA sequence. The misspelled sequence then produced this particular protein, which acts in a way that is slightly different from the original protein. Similar errors in DNA transcription take place all the time as we each reproduce eggs or sperm, but it is extraordinarily rare for such a haphazard change to be preserved and handed down intact for millions of years. The most likely explanation for this is natural selection; in some way, it has worked to the advantage of that hominid species and its descendants to keep both forms of the COMT gene in the family.

The so-called valine and methionine forms of the COMT gene, therefore, offer an example of the evolutionary variation in human ancestry. They also represent a subtle difference in how COMT works in the brain, so that a certain degree of the variation in how well people carry out executive tasks may be attributed to their genetic heritage. Of course, some measure of the variation in cognitive performance depends on the individual's IQ (intelligence quotient, a mathematical formula that attempts to assess mental development and ability), which is itself partly shaped by genetic factors. But most of the variation is due to small, everyday circumstances: how well you slept the night before, whether you have a hangover, or even how interested you are in doing your best that day.

Although genetic factors account for only a small portion of the normal human variation in executive cognitive function, these factors are powerful enough to generate a prediction based on the COMT gene. Every individual inherits two copies of the gene, one from each parent. The prediction is that people whose genotype is met-met (two copies of the methionine form) will, on average, perform the best on specific cognitive tests, and people with the val-val genotype (two copies of the valine form of the gene) will do the least well, while people who carry one valine and one methionine form of the COMT gene will perform somewhere in between.

Experiments carried out with people who have the met-met, met-val, and val-val genotypes have consistently upheld these predictions, as a dozen published studies now attest. COMT genotypes have been calculated to account for 3 to 4 percent of normal human variation in executive cognition and working memory. That may not sound like a great deal, but for one gene out of perhaps 25,000 to explain this much variation in a fundamental feature of human cognition is in fact quite significant.

The COMT gene has also been studied with neuroimaging. This line of inquiry does not study the impact of one genotype or another on cognitive performance but rather asks, Do the genotypes have different effects on how the brain actually handles information? Research by Weinberger and others at the NIMH indicates that COMT acts specifically on the efficiency of information processing—that is, the ability of the neurocircuits to stay attuned to selected signals that carry the critical information and not be distracted by other, irrelevant signals. This observation might hold implications about the ways in which drugs that work on the efficiency of information processing might differentially affect people with different COMT genotypes. One example comes from amphetamines, a class of drugs that increase the neurocircuits' alertness to the selected signals. Weinberger and his team have given an amphetamine or a placebo to normal human subjects and then asked them to carry out certain cognitive tasks. The scientists made predictions beforehand on the basis of each participant's COMT genotype as to how the amphetamine would affect his or her performance.

Clearly, not everyone would respond in the same way, because each has a different genetic background and therefore a different baseline level of dopamine in the cortex. The val-val hypothesis predicts that people who have only the valine forms of the gene will perform better on amphetamines—and indeed, this would not be surprising, given that many people feel that they get a little sharper on amphetamines. But the real test of this genetic prediction is what happens in individuals who have the met-met genotype. These people should do worse on amphetamines, because the drug should take them beyond the optimal level of dopamine for cognitive functioning. And this is exactly what happens. The NIMH study essentially shows that val-val people do better on amphetamines, whereas the cognitive processing of met-met people deteriorates on amphetamines. COMT genotype may not be the only factor that determines the cognitive response to amphetamines, but it demonstrates the existence of genetic factors that affect each individual's response differently.

To put the genetic findings to one more test, the NIMH team targeted this gene and its protein product directly by looking at a COMT inhibitor called tolcapone, which specifically blocks the action of COMT on dopamine levels. In studies involving more than forty research participants, the scientists found that tolcapone enhances normal cognition in many people as the cognitive tasks increase and become more complex, and that val-val subjects are especially likely to show this response.

Neuroimaging studies show that tolcapone raises the efficiency—that is, the signal-to-noise ratio—of executive cognitive processing while people are doing these tasks. Like amphetamines, tolcapone improves cognitive performance by modifying the way the brain handles information. And as with amphetamines, people with the met-met genotype find their cognitive processing to be worsened rather than improved by the drug—demonstrating once again the extent to which genes interact with drugs.

THERAPY VERSUS ENHANCEMENT

Despite the picture that researchers have so painstakingly developed, as described above, of how amphetamine works in the prefrontal cortex of the brain, neuroscientists still disagree among themselves on whether this type of drug enhances learning or simply enhances performance. One reason for this snag is the physiological fact that neither performance nor learning takes place in isolation. Consider, for instance, the drugs known as beta blockers, which are often used by musicians and by golf players to still the normal tremor of the hands. This appears to be a straightforward example of a drug that enhances performance rather than learning. But for a musician or a golfer who practices every day with the help of a beta blocker, does the drug subtly alter the learning process from what it would be with a natural tremor? The distinction between learning and performance may not be as clear as we would like, says Anjan Chatterjee, associate professor of neurology at the University of Pennsylvania.

As another example, amphetamines are sometimes given as an aid to rehabilitation to people who have partial paralysis or are unable to speak after a stroke. The idea, which has gathered some evidence and is still under study, is that the combination of the amphetamines and either physical therapy or speech therapy promotes plasticity—the growth of new connections—in the brain and presumably leads to better recovery. Does that help improve learning, or performance?

In the view of Jonathan Moreno, director of the Center for Biomedical Ethics at the University of Virginia, the distinction between enhancement and therapy follows from our ideas of the distinction between health and disease, which in turn derive from our definition of normality. The philosophical literature of the late 1970s contains various attempts to define normality, and the one that seemed to prevail was the notion that "normal" means, more or less, "typical for that species." This definition opens the door to many challenging examples, such as the disruption of sleep patterns in men after about age fifty. Most physicians would agree that this is a very common condition—one might even say species-typical. If it is, and if a man suffering from this condition were to ask his doctor to prescribe a stimulant to help him stay awake and alert during the day, would he be requesting therapy or enhancement?

Beta blockers, which some studies suggest have the ability to weaken the power of certain memories, may pose a similar question. The research, still in its early stages, is based on the phenomenon in memory called reconsolidation, in which memories evoked by a situational cue somehow become labile again and then are rerecorded in the brain, along with the emotions that accompanied the original event. If the event caused anxiety, distress, and fear, those emotions will be aroused with every memory of it. Treatment with a beta blocker is aimed at lessening the emotional charge of those memories by interfering with reconsolidation.

Moreno describes experiments in which scientists knocked out a gene in mice responsible for both innate and learned fear. The gene is expressed in the amygdala, the brain site in charge of perceiving and responding to threat. Mice with the modified amygdala could not initiate the usual cascade of negative feelings and lost their ability to associate feelings of danger with aversive environments.

It would be unethical to create "knockout soldiers," but similar effects may be achieved in people with a class of drugs that work on the amygdala, preventing the release of stress hormones. Beta blockers have shown benefit in people with posttraumatic stress disorder, such as soldiers who have come home after several years of combat. Whether we see such treatment as therapy or enhancement seems to depend on whether we think the ferocious power of traumatic memories is normal or not. The President's Council on Bioethics debated this topic in its 2003 report, *Beyond Therapy.* In Moreno's view, though, the question of therapy versus enhancement is one of many philosophical distinctions that provide a general orientation but not enough guidance to resolve the matter.

It is a philosophical error, says Murray, of the Hastings Center, to think there may be a single, simple ethics of enhancement. To demonstrate this, he offers the hypothetical example (which may not remain hypothetical much longer) of a drug that neurosurgeons could take to reduce the natural tremor of their hands and enhance their ability to focus. This hypothetical drug has few or no side effects. Multiple studies have found that neurosurgeons who take this drug before an operation produce better results: fewer mistakes in surgery and lower morbidity and mortality rates in their patients. Would it be unethical for a neurosurgeon to take this drug because it provides "enhancement"? Suppose someone we loved needed brain surgery, and two neurosurgeons were available. If one said, "Indeed I take this drug because it helps my patients," and the other said, "Not only do I not take the drug but I actually use nineteenth-century instruments because they better display my technical virtuosity," which surgeon would we all choose? The answer does not require long thought or debate when the life of someone we care about might hang in the balance; an ethics of enhancement would play no part in this choice.

To think that we can find ethical guidance about enhancement in some concept of the natural would also be an error. The President's Council on Bioethics took this approach in its report, suggesting that looking inward to our own nature would allow us to see what is or is not permissible. Murray proposes the opposite instead. We need, he says, "to look outward to the world that we create, to the institutions that shape our societies, and to the relationships, especially the most intimate and enduring relationships in our lives—those with our parents, our partners, and our children—those relationships that are so central to our flourishing—and to ask, 'What will be the likely impact of any particular enhancement technology on the possibility of fulfilling those relationships?'"

Still another error is to expect that we can develop one standard ethical response that will fit in all contexts. In a public discussion, Murray put forth the hypothetical possibility of a drug that could strengthen a person's ability to read other people's emotions accurately while at the same time blocking his ability to share in those emotions. The user of this drug would be in a peculiar state of half-empathy: he would know what others were experiencing but simply would not care about it. In other words, says Murray, the drug would transform him into a high-functioning sociopath. This example serves to underline the crucial point that "enhancement" itself is not an isolated undertaking. When a medical step is taken to enhance a certain attribute or function, others are inevitably affected, whether directly or indirectly.

Today a number of drugs—methylphenidate, amphetamines, modafinil, and the selective serotonin reuptake inhibitors (SSRIs)—are already being used by people who are not ill. In the near future, there will be new agents to enhance wakefulness; quite a few companies are working on this, developing compounds that alter the brain's use of dopamine or of histamine. Other drugs in the pipeline will act on the neurotransmitter glutamate or on the cyclic AMP (cAMP) response element-binding (CREB) cascade—a key protein interaction of memory formation—to enhance memory. Also in development, as mentioned earlier, are drugs to suppress unwanted memories. In addition, chemical agents that suppress anxiety without causing the sedation of the benzodiazepines will soon be available. Although the drugs listed here are all being developed to treat illness, many of them will also be put to use—and will be efficacious—in people who are not ill.

This "off-label" use—that is, the use of drugs to treat ailments other than those for which they are explicitly approved—may bring about a shift in the standard balance between risk and benefit. We would probably be less willing to tolerate side effects of a drug, especially dangerous side effects, if we were well and the drug was optional, so to speak. We might also be less willing to accept a risk, as yet unknown, of long-term harm to our challenged but healthy children.

In order to think clearly about therapy and enhancement, we must remember that both psychotropic drugs and lived experience can produce long-term effects on brain and behavior. Both, in different ways, alter gene expression and lead to synaptic remodeling, causing new dendritic spines to sprout and new receptor sites to bloom. Any intervention, especially in the absence of illness, calls for thoughtful consideration and due respect for the possible long-term consequences.

Chapter 10 ⟿

Government agencies provide major funding for neuroscientific research; the agencies then use the results as they see fit. No neuroscientist working in the United States can afford to ignore this arrangement, which is beginning to raise troubling questions for some researchers who find their work applied for military or strategic purposes about which they may have serious reservations. The debate on this topic overlaps with a knot of questions about treatment, enhancement, and the distinction between them, in which psychiatrists and neuroethicists as well as neuroscientists are striving for consensus.

DUAL-PURPOSE RESEARCH

In the near future, beta blockers may be developed that can not only help treat posttraumatic stress disorder but also protect soldiers in the field from forming traumatic memories that could make them less effective the next time they are in combat. In effect, the soldiers would be "prophylaxed" against trauma. The public has heard very little discussion about this, says bioethicist Moreno. The military use of beta blockers as a pretreatment for trauma would be a bad thing in general if it were to leave a soldier entirely unmoved by bloodshed or death, according to Edmund G. Howe, psychiatrist and bioethicist at the Uniformed Services University of the Health Sciences. On the other hand, from the standpoint of war planners, it might be a good thing for combat soldiers to have a reduced level of fear for the sake of their combat effectiveness.

Moreno explains that many technologies of great interest to neuroscience have a dual-use aspect—that is, the top-quality work being done today by many neuroscientists is of interest to the funding agencies for different reasons than those of the neuroscientists themselves. The funding agencies often are seeking to apply the scientific work in a military or national-security context. The neuroscience world has not really thought about this, says Moreno, but some day soon it must begin to do so, under increasing pressure not only from conspiracy theorists on the Web but also from frightened citizens.

The complex relationship between the behavioral sciences, neuroscience, and the state actually has a long history. The first psychologist in the Office of Strategic Services, which later became the Central Intelligence Agency (CIA), was Henry Murray, perhaps the most distinguished personality psychologist of his time. Murray led the new field of personality psychology in the 1940s and 1950s. His book *Explorations in Personality* (1938) is a classic, and he developed the Thematic Apperception Test for the U.S. Army. This test is still widely used today to help identify dominant drives, conflicts, emotions, and other aspects of personality.

One example of dual-purpose research is the cutting-edge work on robotic prostheses (which will be discussed in Part IV). This research has been sponsored in large part by security agencies. Are they interested in helping amputees? "I'm sure they would like to be able to do that," says Moreno. "Is that their primary mission? In candor, however well meaning the Department of Defense is, its mission is not that of a health agency. They have other goals in mind, including developing robots that could serve in combat, performing mine-clearing and intelligence operations, thereby reducing human casualties." Another example could be memory enhancement, which takes many different forms, from pharmaceutical enhancement to electromagnetic enhancement. The use of fMRI represents further examples, whether for tracking a soldier's brain states while he or she is going through exercises or for screening people for specific military assignments.

Almost everything that neuroscience is working on has the potential to be of military interest. The current dilemma for neuroscientists, Moreno says, has something of a precedent in the moral discussion that preoccupied atomic physicists after the end of World War I: Now that we have developed this supremely powerful technology, how can we make sure it is never again used for harmful purposes? One very important result of that discussion has been the policy of "No first use" of atomic weapons.

But "No first use" may not be an appropriate position for neuroscientists today, says Moreno. He envisions, for example, the futuristic prospect of long-range imaging of individuals in terrorist cells; whether the same imaging technology could or should be used on people in airports would be a different matter. Although the physics of current neuroimaging devices makes this impossible today, other "nonlethal" technologies that might be used in hostage situations, like aerosolized anesthetic agents, are under active study. Would they be acceptable in domestic incidents or only in remote places like Iraq and Afghanistan? It seems the moral questions of neuroscience and neurotechnology are both more complicated and more subtle than those faced by atomic physicists sixty years ago.

WHAT CAN WE DO AND WHAT SHOULD WE DO?

In the vigorous debate over the use of drugs for enhancement, we sometimes overlook the down-to-earth fact that drugs of enhancement do not always offer a competitive advantage. Some do, such as Ritalin in the classroom, but most (for example, an SSRI to brighten mood) do not affect other people significantly. We need not assume that the effects of enhancement drugs would all be negative, although some social concerns do call for discussion. One is the prospect of implicit coercion, which might be felt by the only student in a high-school class who does not sign up for an SAT preparation course, or the only employee in a fast-paced company who chooses not to take a stimulant in order to work eighteen hours per day. Should these pressures be resisted at the individual level or by means of some policy? Should they be resisted at all?

Martha Farah of the University of Pennsylvania points out the basic metaphysical difference between persons and things: people, but not things, deserve credit or blame for their actions. Leon Kass, chairman of the President's Council on Bioethics from 2001 to 2005, has written about the sort of meaning and the dignity that is conferred on our lives by hard work. In fact, a person's sincere effort has intrinsic value, regardless of success. Yet when we change our brains and therefore ourselves by adding new pharmaceutical compounds that alter our neurochemistry, we are in some sense improving our performance the way we would improve the performance of a car, opening the hood and going in and tinkering.

People also have value, independent of how well they perform what they do. We value a car to the extent that it runs well and looks good and takes us where we want to go—in other words, it performs well. But we don't value a spouse or a child because of how well he or she performs. We value them because of some essence of their personhood that we care about—the very essence that we instinctively feel comes under threat of distortion or replacement whenever a medical intervention touches on the brain.

Discussions about using drugs for enhancement sometimes make reference to recent scandals over drug use in professional sports, drawing a contrast between athletic contests and competition at work or school. Chatterjee suggests, however, that the contexts are more similar than different. In the 100-meter dash, for example, the difference between being first and being fourth is minuscule in real-world terms—perhaps several hundredths of a second—but the results of that difference are very great. In sports it is clear that small, incremental changes can have disproportionate, even life-altering, outcomes; the same may be true in many a scholastic contest or on-the-job competition.

Chatterjee offers the following cautionary tale. A hypothetical patient of his, an ambitious executive, is expected to work eighty or one hundred hours per week. The quality of his family life is very poor, and he becomes divorced. He is traumatized by this, and the physician prescribes for him an SSRI, which seems to work reasonably well.

Meanwhile, the daughter of this patient is having trouble in school. The school psychologist makes the judgment that the girl has ADHD, and at the father's request his doctor gives his daughter a prescription for stimulants. His son, in high school, is a talented middle-distance runner but not quite good enough to compete at the state or national level. The father, having read in a medical journal that Viagra increases oxygen-carrying capacity in the lungs (among other effects), asks his doctor to prescribe Viagra for his son in the hope of helping him become a national-class runner, and since his son agrees to this, the physician complies with the request.

Furthermore, the patient, who is vying for a contract in Saudi Arabia, would like to learn Arabic in order to gain an edge over his competitors. He asks his physician, and the physician agrees, to prescribe for him a safe dose of an amphetamine, which he will take shortly before his language lessons to help him learn more efficiently. (Recent evidence from work on language recovery after stroke suggests that administering an amphetamine along with a learning experience tends to make the learning more successful.) He takes his amphetamines and his lessons. When he is ready for his business trip, the doctor gives him his patented travel pack, which is a certain sleeping pill to take when the plane departs and a stimulant to take when it lands. In Saudi Arabia he amazes everyone with his knowledge of Arabic and is awarded the contract.

All is well—unless, of course, we read this all-too-plausible account as a tale of dystopia. In either case, says Chatterjee, this state of affairs is arriving soon—in fact, some aspects of it already exist.

Philosophy has a name for the scenario in which each of us pursues only what is good for ourselves: the tragedy of the commons. It is as if we were all shepherds, with one common grazing area for our sheep. Each of us would like to own more sheep, but every additional sheep provides a benefit only to the shepherd who owns it, while taking a greater toll on the common pasture. It is in the individual interest, but against the common interest, to add sheep; if everyone did so, the pasture would become overgrazed, leading to mass ruin. Unfortunately, in the real-world scenario, it is not at all clear what the commons are.

Ethicists concern themselves with issues of fairness, among other issues; unfortunately, as Steven Hyman has pointed out, it is not clear that the notion of fairness applies often in everyday life. According to Anjan Chatterjee, some high-school students gain an unfair advantage over others by taking an SAT preparation course. In truly disadvantaged parts of the country, however, such as North Philadelphia, where Chatterjee works in a nonprofit, community-based organization, the struggle is to keep people in school through twelfth grade, regardless of whether or not they ever take a college admissions test. "I think these issues are deep," he says, "and there's no reason to think we're going to be able to deal with them in this arena any better than we have in the past."

The most distressing and troubling aspect of this issue is what it does to our notion of personhood. Apparently, a pluralistic society produces little agreement on what personhood might be, except in extreme situations. In an article Chatterjee wrote on this topic, he included the question, "If you had the opportunity to give your child a relatively safe drug half an hour before piano lessons and the child then learned to become a terrific piano player, would you do it or not?" *Popular Science* posted the question on its Web site in an informal survey of its readers. The votes came back almost evenly divided, with 48 percent on one side and 52 on the other.

When Chatterjee later talked informally to some of the respondents, he found they were uniformly surprised by the split vote, because each one was convinced that his or her view of the question would be shared by everyone—indeed, that it was the natural conclusion. Chatterjee suggests, therefore, that on some of the most interesting questions of neuroethics, consensus may be harder to find than we might think.

The possible impact of pharmacology on our ideas of personhood comes up in other contexts besides that of the enhancement of musical aptitude. Martha Farah poses a different hypothetical question: If, after you fall in love with someone who is on Prozac, the person stops taking the drug and you find him or her to be a very temperamental and difficult person, whom did you love? Is it the same person, or do you say, "This is a different person, I don't love this person?" Or is it the same person, whom you love only when he or she is taking Prozac? These issues, says Farah, "are among the most troubling in connection with psychopharmacological enhancement," she says, "because they scramble our nice ontology, which holds that there are persons in this world and there are things, and that we know how to value each and how to deal with each."

As Weinberger sees it, what seems to make the issue of using drugs for enhancement somehow different from many other things our society routinely allows is that drugs could be made available to everyone and that they would be relatively inexpensive. Currently, mainstream society poses no objection to plastic surgery for any newscaster over the age of forty or so. "Maybe we should, but we don't," Weinberger says. "So what is the one thing that makes using a drug different from other means of enhancement? That's the question I'm struggling with."

As researchers, clinicians, legal experts, and ethicists begin to talk about neuroethics and the ethics of enhancement, they cannot avoid confronting these broad questions, says Murray. He points this out not as bad news but as an affirmation that the issues at stake are as complex as they are important. "We need to engage each other and the public and policymakers in a very broad and sustained debate so that we can argue over what would be the most constructive framework within which to talk about enhancement and the neurosciences," he says. "A lot of the fights we have are fights over meaning, over deep social meaning, and I think we need to accept that. We need to have that debate."

"DRUGS IN THE BRAIN" (CHAPTERS 8 THROUGH 10) IS BASED ON PRESENTATIONS BY:

Anjan Chatterjee, Associate Professor of Neurology at the Center for Cognitive Neuroscience, University of Pennsylvania;

Martha Farah, Professor of Psychology and Director of the Center for Cognitive Neuroscience at the University of Pennsylvania;

Gerald Fischbach, Executive Vice President for Health and Biomedical Sciences and Professor of Pharmacology in the Center for Neurobiology and Behavior at the Columbia University College of Physicians and Surgeons;

Ruth Fischbach, Director of the Center for Bioethics at the Columbia University College of Physicians and Surgeons;

Steven Hyman, Provost of Harvard University and Professor of Neurobiology at Harvard Medical School;

Russell Katz, Director of the Division of Neuropharmacological Drug Products at the U.S. Food and Drug Administration;

Jonathan Moreno, Director of the Center for Biomedical Ethics and Professor of Biomedical Ethics at the University of Virginia;

Thomas Murray, President of the Hastings Center;

Benedetto Vitiello, Chief of the Child and Adolescent Treatment and Preventive Interventions Research Branch at the National Institute of Mental Health; and

Daniel Weinberger, Director of the Genes, Cognition, and Psychosis Program at the National Institute of Mental Health.

Part Four

Neurotechnology

Chapter 11 ⌒⟩

The advances in neuroscience, engineering, and computing that made possible today's sophisticated field of neuroimaging have also led the way to numerous other techniques for examining and aiding the brain. "We're at the dawn of a new age of neurotechnology in which a whole host of devices to diagnose and treat nervous-system disorders and to restore lost neurological functions are going to come about, and there's the possibility to augment existing functions as well," says John Donoghue, director of the Brain Science Program at Brown University. To gain a sense of how quickly technology develops, he says, one has only to think of the earliest cardiac pacemakers as they appeared in the 1950s: instruments about the size of a breadbox, carried outside the body and connected to the heart by wires passing through the skin. Nowadays, of course, the standard pacemaker is implanted under the patient's skin on the upper chest and is only the size of a silver dollar.

A NEW AGE OF NEUROTECHNOLOGY

Donoghue attributes the growth spurt in neurotechnology to a great increase in neuroscientific knowledge that has come just as the huge research investments made by the government, public, and private sectors are reaching maturity. These investments, he says, are now beginning to pay off many times over, not only in terms of our understanding of the brain but also in terms of providing significant medical benefits. Yet along with these achievements usually come new ethical questions—some straightforward and others more complex and troubling.

Two kinds of neurotechnology are advancing particularly rapidly: one that transmits or augments signals going into the brain, and another that receives signals produced by the brain, decodes them, and then helps to carry out their original purpose. It is the latter kind that shows promise of being able to restore movement or sensory function that a patient may have lost through brain illness or injury.

Some of these prospects lie in the future, but a few have already been realized; a prime example is the cochlear implant, essentially an electrical stimulation device that goes into the inner ear. Approved by the FDA in 1984, the cochlear implant is now used by more than thirty-seven thousand people in the United States alone. Unexpectedly, the cochlear implant has been found to exert not only a therapeutic effect but a restorative effect: the electrical signals succeed in rescuing cells that would otherwise die off in the absence of stimulation.

A second example, but one that is in a much earlier stage of development at present, is artificial vision. At least two different strategies are being explored, both using video images: either bringing the images inside the eye to stimulate the nerves on-site, or delivering the images directly into the brain itself. An electrode array, invented by William Dobelle, of the Institut Dobelle in Zurich, has stimulated the visual circuits only enough to cause the research participant to "see" a fading dot. The eventual goal of this research is to provide a succession of stimuli that might give rise to a recognizable visual pattern. To differing degrees, the vision-stimulating electrodes and the cochlear implant represent the current prospects for restoring sensory input to the brain.

Neurotechnology to restore movement would work along the same lines. The first stage would require a neural sensor, something to pick up the brain signals related to, say, moving an arm. These signals come out in a very complex form, so the next stage would be developing a way to make sense of them. In the third stage, the decoder would connect to a variety of technologies, perhaps interacting with a computer, so that the patient could hook up to a wheelchair for mobility, to a robot to carry out certain tasks, or even to his own limbs, if they have become unusable through a movement disorder or a spinal cord injury.

The most difficult stage in all this is the first one, capturing the signals that the brain is trying to send to the rest of the body. In theory, these can be picked up either from outside the brain or from within it—that is, extracranially or intracranially. Intracranial techniques can detect sharp spikes of electrical signaling called action potentials, which are a rich source of information about localized brain activity. Extracranial technologies such as electroencephalograms (EEGs) have the advantage of being noninvasive, but the electrical signals they capture are very diffuse, says Donoghue. These report on the brain, reflecting a change in many cells at the same time. They generally have a slow time course, as compared with the speed of thought in the human brain. It is possible, however, to modulate these signals through a kind of biofeedback exercise.

Recording these electrical signals requires putting sensors, that is, disks or electrodes, on the scalp of a willing research participant. Each electrode collects information on the activity taking place beneath it in the brain, and when all the information is put together it yields a smeared, filtered average of regional brain activity. Even so, it is possible to differentiate between the alert state, in which the brain waves are of low amplitude and high frequency, and sleepy states, in which the brain waves are of high amplitude and low frequency. In addition, if the individual undergoes a seizure of some sort, that shows up as a distinct rhythmic pattern in the brain waves. Thus, these signals give a good assessment of the state of the brain. In contrast to action potentials, EEGs that are triggered by a visual flash or an abrupt sound are called event-related potentials, or ERPs.

If an image is flashed before the eyes of a research participant, the response that is recorded by a single electrode, from a single neuron, is an ERP. This one signal will eventually become averaged together with many more that represent the response in this part of the individual's brain to an image that is flashed in front of her—for example, a photograph of her house. Over the course of many trials with many different images, the participant's signaling responses to the image of her own house can show a characteristic wave form.

Some scientists think this ERP response—the 300-millisecond wave form, or P300—may in fact represent the person's familiarity with the stimulus being offered. Whether or not the person wishes to reveal her prior exposure to that stimulus, the P300 wave form will reflect her familiarity with it and so could be seen as a kind of lie detector. However, ERP recording would not lend itself to certain other uses: for example, it would not be ideal as a control signal to allow a patient to move a mechanical device as if he were moving his own arm, because event-related potentials are not a very rich source of signals.

A NOTE OF CAUTION

The distinction between extracranial and intracranial devices is not only technical but also psychological, points out Paul Root Wolpe of the University of Pennsylvania. A great many people, at all levels of education and with all kinds of religious orientation, believe deeply that there is an essence of ourselves called human nature that is somehow disrupted when scientists attempt to augment or replace anything in the brain, but remains intact when we augment, say, the power of the human voice by using a telephone. Most of us feel there is a qualitative difference between picking up the phone to call a friend who lives miles away and having our brain altered so that we could communicate with that friend telepathically. That difference may be ethically relevant or irrelevant, but a scientist who downplays public concerns about technological tinkering with the brain is making a mistake, says Wolpe. Although these concerns may be part of a naturalistic fallacy, people have fundamental beliefs about the integrity of their bodies that modern science ignores at its peril.

Chapter 12

A new technology that uses a small electrical device to deliver signals to the brain holds out hope for relief from the symptoms of Parkinson's disease, other movement disorders, and possibly other chronic brain ailments as well. Many people find this technology troubling to contemplate because it uses electrodes implanted in the brain. The following explanation of how this technology works and just what it can do may contribute to public discussion by clarifying a few basic points.

DEEP BRAIN STIMULATION FOR MOVEMENT DISORDERS

As of mid-2005, approximately thirty thousand people had received implants of deep brain stimulation (DBS) electrodes in their brains to treat a variety of neurological conditions. By far the most common use of DBS technology occurs in the realm of movement disorders, and in particular Parkinson's disease, which accounts for 80 to 90 percent of these patients. However, other movement disorders such as dystonia involve involuntary twisted movements of the body and are also being treated with DBS. According to Andres Lozano, professor of neurosurgery and neuroscience at Toronto Western Research Institute, the treatment of all these disorders has a common theme: to seek out the neurons that are misfiring in the brain and bring them under control.

One striking feature of the disorders in question is that they are caused by dysfunction in a relatively small population of neurons that is then able to interfere with large networks of signaling pathways in the brain. Functional neurosurgery has developed techniques for calming the erratic firing patterns of these neurons, and as the techniques take effect they allow the rest of the brain to function in a more normal fashion.

The first step, of course, is to locate the malfunctioning population; this is often done by MRI. Next, the surgeon introduces an electrode through a hole in the skull and then moves it cautiously through the area while listening to the spontaneous activity of neurons and the rate of neural firing (which sounds like a series of slow and fast clicks on a radio receiver). Neuroscience is beginning to understand the language of the brain, Lozano says, and one important discovery has been that the neurons at different sites in the brain speak different languages, much like people in different countries. By learning these languages, neurosurgeons have been able to navigate through the complex territory of the brain to find the target population.

This procedure applies whether the patient is being treated for Parkinson's disease or another neurological disorder. Once the surgeon has pinpointed the target neurons, she can place electrodes in their midst and connect them to a pacemaker device, which is then programmed to deliver a constant source of current to these targets. Steady current serves to break up the pathological behavior in these localized brain areas, with the aim of influencing larger brain networks and, ultimately, reducing neurological and psychiatric symptoms.

Lozano's team has carried out this procedure on scores of patients with Parkinson's disease. One patient, who suffered from tremor, had electrodes implanted in the subthalamic nucleus of the brain and is now free of tremor as long as a steady current is delivered to the electrodes. Another patient, a former veterinarian, was able to move only very slowly, with a bent posture, before her operation. She had had to stop working years earlier because she was too unsteady to stand for long periods of time. This is a common problem in Parkinson's disease; patients spend part of their day in good condition, but another part in this so-called "off" condition, in which they are rigid and unable to initiate any movements. Three months after the surgery and with bilateral implants in her subthalamic nucleus, this patient was able to stand and to walk when the stimulators were switched on. She has been back at work for three years now under the beneficial effects of deep brain stimulation.

Various research teams are investigating the use of DBS in the treatment of other neurological disorders as well, including intractable generalized epilepsy. Their chosen target is another part of the thalamus called the anterior nucleus. This site is located at a crossroads where signals from the temporal lobes are collected and transmitted to the cingulate lobes and then back to the temporal lobes, so it is a way station for seizures from temporal (memory) and limbic (emotional) networks. Thus, the anterior nucleus appears to be involved in the propagation and generalization of seizures. Intervening at this site could perhaps interrupt the thalamic cortical rhythms and throw off the timing of neuron firing that is important for epileptic seizures. Lozano and his colleagues embarked on a pilot study, implanting electrodes in the anterior thalamus of a small number of patients, with encouraging results. The device manufacturer, Medtronic, is now conducting a somewhat larger study to assess the safety and efficacy of this treatment.

The type of implant under discussion is particularly well suited for scientific studies because its electrodes can be used not only to stimulate a target area but also to record from it. This special opportunity has led to a key observation in a patient who suffered from generalized seizures. Just before a seizure, the electrophysiological activity of the brain issues a kind of warning. A characteristic change in thalamic EEG is now known to precede the onset of a generalized seizure by several seconds, providing a brief interval in which to intervene with electrical stimulation that interrupts the growing rhythmic pattern and thereby prevents the seizure. In Lozano's view, the contingency-like approach of stimulating the site only if and when an electrode detects a warning signal is more effective than stimulating the site continuously. A device of this kind, a so-called loop system with an input structure, a detection event, and a therapeutic stimulation event, is now in development.

HOW DEEP BRAIN STIMULATION WORKS

Despite the clear evidence that delivering an electrical current to specific brain sites through implanted electrodes can bring several beneficial effects, neuroscientists do not all agree on exactly how this happens. Apparently DBS blocks or disrupts the pathological neural activity, but the mechanism through which it does so is unknown. Each possible explanation has its advocates. For example, noted neuroscientist Rodolfo Llinas thinks that when a brain site is stimulated at a high frequency, the synapses are simply overwhelmed and soon begin to fail altogether. Lozano is inclined to think that stimulating certain neurons brings about the release of inhibitory neurotransmitters.

To test his hypothesis, Lozano collaborated on an experiment with a research team headed by Jonathan Dostrovsky, of the Toronto Western Research Institute. Working within a brain site called the globus pallidus, the scientists introduced two minielectrodes simultaneously. This enabled them to stimulate the top electrode while recording from the bottom one and thus to study the effects of stimulation not just on the neuron receiving it but also on those nearby. Their analysis brought to light a train of spontaneous activities in the nerve cells of the globus pallidus. These neurons are extremely busy, firing off signals at the rate of about one hundred per second. But the scientists found that when the top electrode is stimulated, the neighboring neurons stop firing momentarily.

Increasing the frequency of the stimulus causes more and more inhibitory periods, ultimately producing intervals of silence that last 15 or 20 milliseconds—long enough, in fact, to shut down these neurons completely. It appears the extra stimuli—that is, those delivered by the experimental electrodes—cause the neuron receiving them to spill its stores of a neurotransmitter called gamma-aminobutyric acid (GABA), the chemical "off" switch for the other neurotransmitters; GABA then travels across the synapse to neighboring neurons and inhibits their signaling activity. Thus, says Lozano, DBS may work simply by putting into overdrive a system for inhibitory neurotransmission that was already in place.

To explore further the cellular and molecular consequences of stimulation, scientists have introduced electrodes into animal brains; by this means they can bring about electrophysiological changes in the hippocampus. But the hippocampus is famous also as a site for neurogenesis (the creation of new neurons), and so the irresistible scientific question is: Can scientists bring about neurogenesis by means of DBS? Further, is it possible to nudge neurogenesis along specifically by stimulating the thalamus?

Clearly, the answer to both questions is yes. After 100-hertz stimulation of the thalamus, samples of tissue from the hippocampus of a rat brain shows approximately double the usual amount of neurogenesis—quantifiable proof of the molecular effects of stimulation. In the future, DBS may be used not only to fine-tune signaling pathways that have gone awry, but also—an awe-inspiring prospect—actually to repair damaged areas of the brain. The ethical pros, cons, qualifications, and implications of such an ability would require the broadest possible debate.

However, it is important to keep in mind that current treatments are only symptomatic. So far, for someone who has Parkinson's disease or one of several other neurological disorders, DBS can do a good job of relieving the symptoms but cannot alter the natural course of the illness itself.

DEEP BRAIN STIMULATION FOR DEPRESSION

With Parkinson's disease in the vanguard, DBS is now under intense study for possible use in treating other disorders. Some trials now under way involve DBS for the treatment of epilepsy, others for the treatment of chronic pain. The greatest opportunity, and perhaps the one most fraught with ethical questions, is the possibility of applying DBS technology toward the treatment of psychiatric disorders.

A survey by the World Health Organization in 2001 found that about 450 million people in the world are affected with a mental, neurological, or behavioral disorder at any given time. Current studies are exploring the use of DBS to treat obsessive-compulsive disorder, Tourette's syndrome, and depression. Still in a preliminary stage are inquiries about DBS for people who suffer from drug dependency and other disorders.

On a worldwide scale, depression is one of the most serious psychiatric challenges. It affects about 120 million people in the world and is the leading cause of disability among workers. Of the estimated 870,000 suicides per year worldwide, a majority are thought to be related to refractory depression.

Psychiatry, the social sciences, and alternative medicine now offer many effective treatments for depression, ranging from medication to cognitive therapy to herbal remedies. Nevertheless, relapses and recurrences are common. About 10 percent of patients feel no benefit from any of their treatments, and this group carries an especially high risk of suicide. New ways to treat refractory depression are urgently needed.

One promising approach not only is new itself but owes its origin to the relatively new field of neuroimaging. Imaging studies by Helen Mayberg, at Emory University, and others have revealed that one site in the brain becomes particularly active during events or moments of acute sadness. The finding suggests that this site, known simply as

Area 25, functions as the brain's "sadness center." Mayberg has also observed that when a patient is treated with antidepressants and his depression subsides, Area 25 in his brain begins to show a lower level of activity than before.

The response of Mayberg, Lozano, and their colleagues to these observations was, of course, to ask whether DBS itself could reduce the level of activity in Area 25, and how this might affect depression. The scientists set out to design a study to address these questions—a daunting task because of the intricate logistical and ethical challenges it presented. In its final form, the study involved ten participants. The number of participants is small because the study called for strict parameters: these were people who had been severely depressed for the preceding twelve months or more and had found no relief in any treatment thus far. They had to have tried electroshock therapy and at least four different regimens of antidepressant drugs, at adequate dosages, for weeks at a time. These were patients with disabling, apparently intractable depression—unable to leave the house and incapable of experiencing pleasure in anything. Yet, at the time of the study, they had to have no suicidal thinking and to have made no suicide attempts in the preceding three years, and they had to be free of other psychiatric ailments.

The first procedure in the study was to place two electrodes in Area 25, one on the left and one on the right, in a brain region that is known anatomically as the subgenual cingulate cortex. Recordings from the electrodes showed at once that Area 25 was hyperactive, whereas other frontal areas of the brain that normally engage in executive functioning were underactive as compared with the normal brain.

When the scientists turned on the electrodes, they were able to reduce the activity in Area 25 to normal and even below-normal levels, and to keep the area underactive for weeks. Meanwhile, the other frontal areas that had been underactive became reengaged and metabolically active. The results for some patients were remarkable. One woman who had described herself before the study as suffering from a sense of nothingness, of "being in a black hole," recovered so well that she not only went back to work but succeeded in starting her own business. Of the first six patients to receive DBS, four improved substantially and two did not. A seventh patient whose surgery was conducted more recently has shown good results as of this writing.

A major challenge for DBS treatment is to account for the one-third rate of failure in Mayberg's study and to devise ways to reduce it. Lozano and Mayberg hope also to expand their study and, in particular, to take advantage of the adjustable nature of DBS to observe the effects of stimulation on an ongoing basis. In one patient, in whom the stimulation was turned off and turned back on again, the results were closely synchronized with DBS activity. Without stimulation the depression returned, but it faded again when the stimulation was turned back on.

ETHICAL AND PRACTICAL CONCERNS
OF DEEP BRAIN STIMULATION

Introducing electrodes into the brain of a conscious human being calls for boldness, precision, and a judicious use of PET scans. Neuroimaging directs the surgeon to the correct area of the brain; then, right in the operating room, the electrodes can be switched on, producing immediate effects in the patient's brain. In an odd variation on the usual surgical procedure elsewhere in the body, it is the patient who guides the surgeon's hand, by reporting on his sensations as the electrodes stimulate first one neuron and then another. "The patient will say, 'Who turned on the lights?' or 'All of a sudden the room's in color,' or 'I feel a sense of calmness,' so we have an acute effect that guides the placement of the electrodes," Lozano says.

DBS is subject to an approval process—several, in fact. The stimulating device itself and all its constituents (the polymers, metals, and so on) must first be approved for safety by toxicological studies. Next comes a so-called pivotal study to show the procedure is safe and efficacious for its intended purpose. Only after clearing both hurdles does a proposed use of deep brain stimulation receive approval from the FDA.

Nevertheless, says Mahlon DeLong, director of the Emory University Functional Surgery Program, however safe and precise a medical intervention into the brain may be, its effects can never truly be highly restricted or local, because every site in the brain is part of a larger signaling circuit. Indeed, many of the disorders that have come up for discussion in these pages are not really examples of localized function but of circuit function or dysfunction. When medical science intervenes at one location, it disrupts activity throughout pathways and networks that are poorly understood even today.

But neuroscientists have begun to piece together how some of these cortical and subcortical pathways function. One scheme proposed by DeLong and his colleagues is that the cerebral cortex, the basal ganglia, and the thalamus are linked by a family of circuits that group similar functions together. These circuits are not the haphazard mass of signaling sites they appear to be but a well-organized system that has developed, broadly, to serve four kinds of functioning: motor, oculomotor (that is, eye movements), associative, and emotional. These are, of course, only the most general categories of much more specific subcircuits that can be identified as well.

As just one example of how these circuits connect with other circuits and seemingly disparate brain sites, stimulation of the subthalamic nucleus (discussed earlier as a treatment for Parkinsonian tremor) may in some cases cause patients to experience considerable emotional disturbances such as hypomania or even profound depression—an effect that can only be explained by the engagement of these other circuits in the brain. DBS has enormous potential for alleviating neurological symptoms, but at the same time it is not without risk, in part because of its invasive nature and because of unintended side effects.

In addition, says Lozano, this technology brings out a number of ethical issues. Obviously, great care must be taken with the application of DBS and other neurotechnology. There are also issues of distribution: these therapies are very expensive, and even apart from the cost, not everyone has equal access to them in the United States, let alone in other countries.

The epidemic rates of depression today illustrate these issues all too clearly. Clinical depression, whether diagnosed or not, affects a disproportionate number of people who are poor, socially disadvantaged, and/or in prison. Mary Faith Marshall, professor at the Center for Bioethics at the University of Minnesota Medical School, asks, Do these special populations have access to new developments in the study and treatment of depression? Should they—and if so, by what means?

Chapter 13 ~~

To say that the most innovative areas in neurotechnology present the greatest ethical problems would be making too simple a formula out of a complex and constantly shifting state of affairs. Yet it is clear that as more fields of expertise come to contribute to medical-scientific initiatives—from neurology, psychiatry, and pharmacology to surgery, genetics, computational neuroscience, bioengineering, and information technology—more situations arise in which the best interests of the patient or research subject may clash with those of the research sponsor or collaborator. Therefore, all the experts involved must be on the lookout for potential ethical challenges and be prepared to address them as an integral part of their work.

THE BRAIN-COMPUTER INTERFACE

Decoding the electrical signals of the brain is for some neuroscientists the work of a lifetime. As for the purpose of such an endeavor, one very appealing goal is the prospect of developing a neuromotor prosthesis, otherwise known as a brain-computer interface. The aim is to help individuals who have become paralyzed to regain an ability to move or communicate or interact with the world. If such technology became widely available it could change the lives of hundreds of thousands of people who are currently unable to move as a consequence of spinal cord injury, neurological disorder, or limb amputation.

What characterizes many of these conditions is that the people affected have a healthy, well-functioning brain. Their central nervous system can issue movement commands, but the commands do not reach their destination. A clear example is spinal cord injury, in which the signal pathway from the brain to the spinal cord has been severed, so that the signals cannot travel to the appropriate muscles of the body. In such a case, the brain-computer interface would serve as an alternate communication system that could pick up the signals from the motor cortex of the brain, carry them outside the nervous system using physical connections rather than neurological ones, and then deliver them to the parts of the body that would act on them. Thus, for the first time in history, it would be possible for neuroscientists to communicate directly with the brain, bypassing the peripheral system entirely.

As discussed earlier, the brain sends its messages in the form of electrical spikes, each about a thousandth of a second long. The specifics of each message are encoded in the number of spikes the brain site is firing off in a particular amount of time. For example, a recording from a neuron in the brain of a person who is thinking about moving his hand to the left or to the right might demonstrate that the neuron fires seven spikes just before he moves his hand to the left and two spikes before he moves it to the right. The code in that case would be quite simple: seven spikes means "left," two spikes means "right." If someone were to record from the same neuron another time, she could say, "I hear seven spikes; I predict this person will now move his hand to the left." This feat might appear almost magical to the uninformed observer, but in fact it is based on the sound scientific principle that messages from the brain are encoded in electrical spikes.

Unfortunately, says Donoghue, in reality the code is not so simple. In any given recording, the number of spikes varies considerably; sometimes it might even be the same for movement to the right as for movement to the left. Only by averaging together the signals of many different neurons from the same brain site can scientists begin to see the real patterns of correlation between the number of spikes and the intended action. As for the variations, at present poorly understood and often dismissed as noise, they may very well hold the key to even more complex signaling systems.

Electrical spikes "work at the speed of thought because fundamentally they are the essence of it," Donoghue observes. The tools for studying these subtle, evanescent signals must meet stringent requirements. For one thing, implanting the electrodes to detect the signals is a delicate operation, as discussed earlier. For another, sensors that can pick up signals from multiple neurons have been difficult to produce, and making them in long-lasting form has been especially difficult. Recent attempts, however, have come close to success—that is, to being workable, though still far from perfect.

Donoghue and his colleagues have created a brain-computer interface for testing in quadriplegics through an FDA-approved trial. In its current form, the system consists of an array of electrodes that are placed on the patient's cerebral cortex, with a connection that comes out through the skin and passes on to a series of computers; the computers decode the brain's intention—for instance, to move the right hand to a particular location—and transform it into the motion of a cursor on a monitor. Thus, the brain's movement commands are translated into the movement of the cursor.

Before making the cursor into a control signal for a variety of devices, the research team needed to determine how well the patient could control it with his brain signals. Accordingly, they designed for the patient a computer program for drawing and asked him to use his neural control of the cursor to make a circle. He quickly mastered this task, as well as the task of using an "eraser" icon to delete a line he had drawn. In addition, the computer interface has enabled him to play a computer game called Pong and to control his television set.

These options, unimportant in themselves, hold great significance for the patient because they represent the beginning of his ability to control his own environment—an ability that most of us take for granted every day. Beyond the patient's immediate environment, though, it would obviously be valuable to him to be able to control devices in the outside world. This highly motivated patient mastered the use of both a robotic hand and a robotic arm in just two demonstrations. In a sense, he did not have to "learn" anything; within minutes of being shown the brain-computer interface, he was able to use the motor cortex of his brain to drive the devices. When the robotic hand, a motor-driven prosthesis, is connected to the neural output, he can open and close the hand at will. He does not need feedback from the computer but can look directly at the device while he uses it. To help him direct the robotic arm, the patient can look at the interfacing device that moves the computerized arm around. By means of just the normal brain signals for moving one's arm, he is able to pick up a small object, transport it from one place to another, and then drop it into someone else's hand.

This device is still at a pilot stage and has been tested on only three patients so far. In order for it to become practical it will have to be made wireless, implantable, and able to run itself, and work is proceeding toward that end.

Ultimately, one very ambitious but possible goal, says Donoghue, is the actual physical repair of the nervous system in people who are paralyzed. This could be accomplished by picking up signals from the motor cortex, conveying them to the muscles, and activating those muscles through electrical stimulation devices. Indeed, such devices already exist, but they are controlled by buttons instead of by brain signals.

If a person's paralysis is so extensive that she can move only her tongue, then a tongue-activated switch will, of necessity, become her mode of interaction with the world. But the aim of the brain-computer interface is not to produce a surrogate but actually to replace the lost function. Whereas the number of things the patient could accomplish with a tongue-activated switch is limited—for example, she would not be able to speak while doing anything else—the limits of the brain-computer interface are not yet known. In theory, they should be the same as the limits of the nervous system itself. Support for this kind of technological development has come from many sources, including the National Institutes of Health (NIH), U.S. Defense Advanced Research Projects Agency (DARPA), Veterans Administration, venture capital, and the public market, says Donoghue. He notes, too, that most of this work has been based on research conducted on nonhuman primates. This raises ethical questions of its own, but about the value of the research there is little debate.

With the rapid development of this technology comes the need for a thoughtful plan of use, says Dennis Spencer, chairman of neurosurgery at Yale Medical School. Local, societal, and medical panels could help health care professionals and consumers to sort out not only how these devices work and interact but how they are distributed and who pays for them. An international forum for just such issues, based at the University of Genoa, Italy, operates online at www.roboethics.com.

As mentioned earlier, electrical stimulation offers hope for the treatment of epilepsy, and a number of clinical trials of this approach are currently under way. Robert Goodman, associate professor of clinical neurological surgery at the Columbia University College of Physicians and Surgeons, is involved in a study of a device similar to one that stimulates the vagal nerve, or tenth cranial nerve. The idea, based on limited evidence from patients in previous trials, is to implant a device that would continuously monitor the patient's EEG with two electrodes, using an algorithm to detect the onset of a seizure and delivering stimulation to abort it. To be workable, this study must be a collaborative effort among multiple medical centers, which will bring extra scrutiny to the selection of appropriate participants as well as to the results.

ETHICS OF NEUROSURGERY

When a proposed new technique entails surgery, one unavoidable ethical issue is whether placebo, or "sham," surgery is required in order to test the effects of the real treatment against something as nearly comparable as possible. Trials of a new drug usually include giving some participants a placebo pill or liquid or powder that looks exactly like the real thing, but the idea of placebo surgery makes many people uncomfortable. As Spencer points out, such operations would be done solely as an experimental control in the process of trying to hasten scientific understanding of whether a procedure was effective—and it could not be guaranteed harmless, as could a placebo pill. In the case of any progressive disease, the simple passage of time should reveal the effectiveness of the surgery as compared with nonsurgical treatment. Nevertheless, in several trials of surgery for Parkinson's disease, the sham procedure was used for comparison, touching off a debate that still continues in several neuroscientific societies.

Ruth Macklin, of the Albert Einstein College of Medicine, observes that sometimes the best principles of science are not consistent with the best principles of ethics. Where does this leave placebo surgery? In Goodman's view, the reason that placebo surgery took place was probably that investigators were eager to carry out a definitive study and wanted to reduce the preliminary investigative work. The answer would have come out through a rigorous analysis of the results, he says, even without a placebo control.

Spencer rejects the proposition that all trials require a placebo and says that when they are necessary, they should be made very specific both to the disease being treated and to the technique being studied. Lozano offers a different approach: "If we're going to do sham surgery, it should be designed in such a way that it is a stepping stone toward developing the definitive therapy," he says. "So, for example, we can make a burr hole in the skull with the understanding that we can use the same burr hole again later, so that this would reduce the prospect of risk and give at least part of the benefit of surgery." Moreover, he thinks techniques have now advanced to the point that everything neurosurgeons used to treat by ablation—that is, surgical removal—can now be successfully addressed with electrical stimulation. Since the latter is more likely to be palatable to the patient and is preferable on both medical and ethical grounds, Lozano predicts that simulation will gradually replace ablation as the standard treatment for a number of disorders.

DeLong points out, however, that there will always be cases in which ablation is necessary. In many parts of the underdeveloped world, access to DBS is almost impossible for economic reasons, and ablation is the only alternative.

The technique of vagal-nerve stimulations carries its own set of ethical problems. One medical-device company was founded solely on the basis of this one procedure; the company created the device, tested it, and was able to show that in certain circumstances it brings a significant benefit for patients with epilepsy. Proving that a surgical treatment, involving an implant, is both safe and efficacious does require a major investment and properly controlled research. But once the procedure—and hence, in this case, the device—overcomes these hurdles and receives approval for specific indications, then, like a drug, it can be applied more widely, and not necessarily under the same circumstances as those for which it was approved. The temptation to find more uses for it is likely to be great—especially, perhaps, for a company whose only product is this device.

DeLong speaks for a number of neuroscientists who have watched with concern as the vagal-nerve stimulator, since its approval, has pushed beyond the boundaries of established epilepsy centers, where its use was carefully considered by experienced specialists. The device is now being promoted to physicians in community practice, including neurologists without experience in surgery. Although it was originally approved for use in patients who could not undergo surgery or preferred not to do so, in many cases now patients are being offered this option and are having the device implanted without either of those two indications. The vagal-nerve stimulator is even under study as a means to treat depression. "When you have a device that's approved for one indication, you're anxious to find other useful indications for it," says DeLong.

Moreover, as Goodman points out, the maintenance of these electronic devices is usually performed by nonsurgeons, and therein lies the potential for abuse. Physicians who recommend the implant may have a financial incentive to do so, because they are the ones who will maintain it with regular, well-remunerated office visits.

Functional neurosurgery, or surgical treatment for neurological diseases, addresses conditions that traditionally have not been treated with surgery; in other words, these patients are normally managed by nonsurgeons. The physicians who take care of them have experience with nonsurgical methods. The impetus for treating them by means of surgery has come from outside, mainly from teams of surgeons working with neurologists. In fact, neurologists have made up the majority of clinicians advocating this team approach. Neurologists who work with surgeons to treat epilepsy gain an appreciation for the role of surgery that they would never develop otherwise. According to DeLong, only these teams are able to judge when surgery is the right intervention for a particular patient.

An early example of such a collaboration comes from a 1989 consensus conference at the NIH on epilepsy surgery. A panel reviewed the evidence for and against the use of surgery and concluded that enough experience had accumulated to show that surgery clearly was an effective treatment for a large segment of patients with epilepsy. The panel also found that epilepsy surgery at that time was underused. This was due partly to a widespread reluctance to use brain surgery, a reluctance that still persists and is useful in one way: it acts as a natural barrier against overuse. But the barrier can be overcome if the patient is desperate; physicians, too, can become desperate and may be willing to use treatments that are still in the process of being proven.

Direct brain treatments such as electrical stimulation raise similar issues, says DeLong. Here again the team approach protects against ill-considered use of the technology, because when the various physicians are obliged to work in a team, discussing whether a patient is an appropriate candidate for surgery, each physician has to convince the others that he has followed the appropriate criteria to recommend for or against the procedure, and they all learn from one another's experience.

Chapter 14 ⟿

As discussed earlier, surgery is not the only aspect of neuromedicine in which the practitioners have an obligation to be sensitive to the ethical implications of their efforts. In a host of other contexts as well, the ethical awareness of the clinician or scientist may be the patient's only shield from various kinds of exploitation.

BUSINESS CONSIDERATIONS

In many fields of medicine today, it is entirely possible for someone to be a clinician, part of a medical company, and a scientific investigator all at the same time. Like many of her colleagues, Marshall worries about the potential for abuse when one individual wears all three of those hats, at least in settings that do not have very stringent oversight. The physician's self-disclosure and the patient's informed consent do not appear to cover the issue fully. If a prospective research participant is informed that the person who will perform the surgery also invented the device, owns the biotech company, or has a major share of the company's stock, does that put the doctor beyond suspicion of a possible conflict of interest? Marshall considers that research institutions and the investigators themselves should have the responsibility of making sure there are procedural safeguards in place.

A desperately ill patient often is willing to try any new approach that offers hope, and may not have either the knowledge or the detachment to determine which treatment is actually best for her. Marshall offers as an example the use of a brain shunt for people who have a disorder known as normal pressure hydrocephalus, a fairly rare condition. A few medical companies now market this treatment directly to patients as a way to treat some unrelated illness such as Alzheimer's disease. People who are susceptible to this marketing may or may not actually have normal pressure hydrocephalus, and they are looking for brain surgery to help them when they are not necessarily appropriate candidates for surgery.

THE THERAPY-ENHANCEMENT DISTINCTION

On first consideration, the difference between therapy and enhancement seems quite clear: therapy is supposed to treat a deficiency or a disorder, whereas enhancement improves an ability or a characteristic. Therapy has the goal of restoring normal health; enhancement aims to make people better than well, to move them from standard to peak performance.

However, says Marshall, sometimes the distinction is not so clear, for several reasons. One is that as individuals we each start from a different point on the spectrum of talents, so that a procedure that might be a therapy for one person is an enhancement for another. Another reason is that each person's abilities change with time, so that what would be a therapy for someone now might be an enhancement for him in the future. Some procedures, such as a method to improve memory or hearing or vision, might even bring about a treatment and an enhancement effect at the same time in the same individual.

Why do most people accept the augmentation of our faculties on the outside of the skull, comfortably wearing glasses or contact lenses or even cochlear implants, yet feel uneasy at the prospect of someone tinkering with the equipment inside? In Marshall's view, this barrier has to do with a shortfall in the public understanding of science, particularly bioscience. A similar uneasiness was evident in early discussions of the human genome project and of genetically engineered foods, she says: "It's a control issue and a fear that at some point scientists are going to unalterably change the fundamental sense of what it means to be human or to control one's world." She notes also a lingering mistrust of the biomedical establishment in this country that goes back to an infamous study conducted by the Public Health Service at the Tuskegee Institute, in Tuskegee, Alabama, from the 1930s to the early 1970s, in which hundreds of black men who had syphilis (which can lead to neurological dysfunction and then death) were left untreated in order to allow scientists to study the natural course of the disease.

Science reporting in the mass media, with its preference for single-issue stories with simple solutions, may further erode the public's scientific literacy. It may now be part of a researcher's job, Marshall says, to help educate the public about the potential uses and abuses of new research. Risk-benefit analysis should also be part of the discussion; in general, the risks that people are willing to accept for therapy are somewhat different from those they might consider acceptable for the sake of enhancement.

In the view of Paul Root Wolpe, the ethical duties of the researcher do not lie solely in his dealings with the media. Each scientist bears part of the responsibility for the conduct of research in general. How do the collective actions of individuals add up to a social set of trends that the great majority of neuroscientists oppose but do not know how to prevent? He says, "We have to ask ourselves not only the microquestion, 'Am I doing this particular study right and am I analyzing it right?' but also 'How do my particular technology and my technique contribute to a greater trend that all of us abhor and yet none of us seems to know how to derail?'"

Each step in neuroscientific research takes the whole enterprise in one direction or another. If researchers wish to avoid seeing their

field move in a direction they do not approve, they must constantly be of two minds. As they engage in their individual projects and
ponder specific neuroethical questions, they must also keep an eye
on how they are contributing to a society-wide set of ideas about
medical science and health care in general.

One question on which neuroethics has much to say is the potential for abuse of brain-computer interfaces, especially when the
applications go from the context of therapy to the context of enhancement. According to Marshall, some of the scientific work now
in progress and likely to be funded in the future—whether by DAR-
PA, the Department of Defense, or the Office of Naval Research—
may have no prospects for use in therapy or in enhancement, in the
usual sense of the word. She recommends, therefore, that military-
funded research should be as transparent as possible, asking, "Can
we monitor and channel it and control it for the common good as
opposed to the commercial interests of a few?"

THE ROLE OF THE NEUROETHICIST

Over the last thirty years of medical research, the prevailing atti-
tude has shifted among professionals from a protectionist stance, especially toward prisoners and children and those who are most vulnerable, to an inclusive stance that welcomes more diverse groups of
participants, patients' advocates, and community spokesmen. Much
of the energy for this shift came from the courageous men and wom-
en who volunteered for human immunodeficiency virus (HIV) research and from the actions of communities with acquired immunodeficiency syndrome (AIDS).

One particularly difficult issue in cutting-edge research is the decision-making capacity of those who may participate. The standard
position among biomedical ethicists is always to involve the patients
or research participants to the full extent of their capacity, whether
or not they are competent in the legal sense of the word. But how

does that play out in practice, especially when an individual is desperate for the only possible cure that exists at the moment or the only chance she may perceive she has? And what happens when a surrogate must make a decision for someone who is not capable of deciding for himself in the forty-nine states that have no law allowing surrogate decision making for research?

Shifting the norm in therapy or in enhancement presumes there was a norm to begin with. But what is normal and what does that mean? Asking such questions is not a bad thing, says Marshall. Rather, it is an acknowledgment that as medical technology progresses, the norm is a moving target. And, to take an optimistic view of human nature, ethics evolve as well; certainly the ethics of biomedical research do so. One point that has become more nuanced is that within the research context, the participants' best interests cannot always be held at the same value as the search for new knowledge. To help address such problems, *The Belmont Report: Ethical Principles and Guidelines for the Protection of Human Subjects and Research,* published by the U.S. Department of Health, Education, and Welfare in 1979, identifies a few key ethical considerations. *The Belmont Report* highlights three basic principles: respect for persons, beneficence, and justice.

"We know what respect is," says William Heetderks, of NIH. "We also know that in many cases, there has not been respect." *The Belmont Report* identifies two components of this principle. One is that individuals should be treated as autonomous agents—that is, the health care professional should treat the patient as if he were capable of deliberating and of making decisions. The other is the importance of recognizing that some individuals actually are not autonomous and that those with limited capacity ought to receive protection. Because it is difficult in some cases to know whether an individual is in fact capable of understanding all the issues brought before him and of deciding on them, Heetderks suggests that each patient be treated not only as an autonomous agent but also as a member of a community. Decisions about treatment and about participation in research would thus not rest solely with the physician or the patient, but would be made in the context of the community.

Beneficence also has two components, according to *The Belmont Report.* "I think every physician hears 'Do no harm' as a first principle," says Heetderks. But he feels it cannot be a guiding principle, because almost any medical procedure does in fact have the potential to cause significant harm.

Therefore, physicians cannot really achieve the ideal level of beneficence; but they can carry out the second component of beneficence, which is to maximize the benefit and minimize the harm. They also have a responsibility to continue working toward the ideal of "do no harm." For a procedure that has a 1 percent risk of complications, he says, if there is research that could cut down the risk to .001 percent or to .01 percent or even just to .1 percent, we as a society have a responsibility to carry out that research.

As for the third principle, justice in the context of research, the issues are who pays for it (by taking the toll of discomfort and possible health risks of participating in a study) and who receives the benefits of it. These are questions that deserve to be addressed at this time in this country, Heetderks says. They are related to justice in the sense that as the medical and scientific enterprise forges ahead with new therapies and technologies, it also carries a responsibility to develop them in such a way that they can be available to all. The only way to bring that about, in his view, is by dramatically cutting the costs of these therapies. In other words, the principle of justice potentially commits us to research into making them widely affordable.

Ruth Fischbach, director of the Center for Bioethics at the Columbia University College of Physicians and Surgeons, points out that although each of the three Belmont principles clearly has great ethical value, in real-life situations they often clash, and then it is unclear which is the overriding principle. Is autonomy more important than beneficence? Does justice take precedence over autonomy? The answers may vary from one case to another and in any case are far from obvious, but neuroethics can at least serve as the common ground on which everyone who has a stake in these questions can come together to thrash them out. Says Marshall, "What we bioethicists bring to the table is a problem-solving method and a body of knowledge in our discipline. We can think about problems in terms of theory, but we can also use our knowledge in an applied way by consulting and by carrying out our own research. I hope we are bringing some critical thinking skills in a prospective way so that we help all concerned to sort through the risks and the benefits, in terms of both individuals and society, of a proposed medical step— ideally, before the work of others in the biological realm is actually applied."

"NEUROTECHNOLOGY" (CHAPTERS 11 THROUGH 14) IS BASED ON PRESENTATIONS BY:

Mahlon DeLong, Director of the Emory University Functional Surgery Program;

John Donoghue, Director of the Brain Science Program at Brown University;

Gerald Fischbach, Executive Vice President for Health and Biomedical Sciences and Professor of Pharmacology in the Center for Neurobiology and Behavior at the Columbia University College of Physicians and Surgeons;

Ruth Fischbach, Director of the Center for Bioethics at the Columbia University College of Physicians and Surgeons;

Robert Goodman, Associate Professor of Neurological Surgery at the Columbia University College of Physicians and Surgeons;

William Heetderks, Associate Director of Extramural Science Programs at the National Institute of Biomedical Imaging and Bioengineering;

Andres Lozano, Professor of Neurosurgery and Neuroscience at the Toronto Western Research Institute;

Mary Faith Marshall, Associate Dean for Professional Conduct and Humanities at the University of Minnesota College of Medicine;

Dennis Spencer, Professor and Chairman of Neurosurgery at the Yale
 University School of Medicine; and

Paul Root Wolpe, Professor of Psychiatry, Medical Ethics, and Sociology
 and Director of the Program in Psychiatry and Ethics at the University
 of Pennsylvania School of Medicine.

Hard Science, Hard Choices

A Public Discussion of Neuroscience, Ethics, and Law

DR. GIFFORD: Good afternoon. I'm Prosser Gifford, the Director of Scholarly Programs here at the Library, and it's my great pleasure to introduce this panel this afternoon. It's very dangerous to introduce somebody who's an expert on language and a longtime political correspondent in Washington, but my job is very simple. I'm introducing William Safire, who will in turn introduce our other two panelists. As you know, since 1979, he has been writing a column on language in the Sunday Magazine Section of the *New York Times* for which he has become very well known. He's now chairman of the Dana Foundation, and it is through the Dana Foundation, along with the Library of Congress, Columbia University College of Physicians and Surgeons, and the National Institute of Mental Health, that we're having this conference. The Dana Foundation supports brain science, immunology, and arts education. Mr. Safire is a 1978 Pulitzer Prize winner for distinguished commentary and the author of fourteen books. So I turn it over to him.

(Applause)

MR. SAFIRE: We're here today with two of the giants in the field of neuroethics. I first met them when we launched the field three years ago in San Francisco, where the University of California–San Francisco and Stanford got together. I thought I coined the word "neuro-ethics," and I was very proud of it until I got a letter from a Danish neuroscientist proving that he got it a few months earlier, so that was a disappointment.

What we have now is Michael Gazzaniga, professor of neuroscience and director of the Center for Cognitive Neuroscience at Dartmouth College, whose career has been spent learning how the mind emerges from the brain. He has a book that's just come out, *The Ethical Brain*, and Mike, as you'll see, is the father of cognitive neuroscience.

Our other conversant is Hank Greely. He is a professor of law and of genetics, which I think is the only mixture of that kind in the country, at Stanford University. In the *Wall Street Journal*, Hank Greely was quoted recently in a story about the question of whether animals and humans should be mixed in any way, and he came up with a stunning quotation which has flown all over the world: "The centaur has left the barn." Now, I'm a language maven, and I could just put myself in his mind. It was either taken from "The horse is out of the barn" or "Elvis has left the building."

(Laughter)

MR. SAFIRE: Let's begin, then, with the notion of the ethical dimension to the chimera. Hank, explain what a chimera is and how you feel about the implantation of animal cells in the human brain or vice versa.

MR. GREELY: There are a lot of different kinds of chimeras. Lots of us may actually be chimeras in the sense that there may be cells of other humans living inside us. Those of you who are women who have had children may well have some of those children's cells that sneaked across the placenta and made it into your bloodstream; they may still be there. Some of us had twins that we didn't know about; the twins disappeared, but left their marks in us. So this mixture of different organisms isn't that new, nor is the mixture by scientists of organisms of different species. Scientists have used chimeras for a long time to study human cells in vivo.

Now, humans are terrible experimental animals. We can't slice us up and dice us up. We're expensive and we make trouble. Mice are much handier, but if you want to study human cells, you'd really like to study them inside a living being and not just in a test tube or a petri dish.

We have long put human cells into mice to grow cancers, to grow other things. In one case, two scientists created something called the Skinhoj mouse, a mouse with a completely human immune system, to study the system in vivo, inside a living organism, but one that you could cut up, unlike studying it inside humans.

The experiment that you're asking about hasn't been done yet. It was proposed several years ago by one of my Stanford colleagues, Dr. Irv Weismann, who was one of the co-inventors of the Skinhoj mouse, to take a strain of mouse that naturally loses all of its neurons shortly before birth and to inject into that mouse in its fetal condition human brain stem cells, in the hopes that as the mouse's neurons died off, the human brain stem cells would turn into neurons and would take the place of the mouse neurons. If everything worked right, which I think is highly unlikely, he would get a living mouse that had a brain made up of human neurons.

Now, I'm not very worried about putting my immune system in a mouse, but we care a little bit more about brains and gonads than about gall bladders and immune systems and pancreases, and he recognized this might be a hot topic. He asked me several years ago to put together a working group to advise him on the ethical issues, and our conclusion was that there is an ethical concern about imparting any human cognitive or mental abilities into a nonhuman animal.

We didn't resolve whether that's a bad thing, good thing, or in between, but we certainly thought it was an appropriate concern. We also concluded his experiment was very unlikely to work, given the tiny size of the mouse brain compared to the human brain and its differences in structure. But we said, "You can go ahead if you do it in the staged way: implant the human cells in the fetus, kill the mice before birth, look at their brains, and look at the structures in their brains."

Mice have specialized nerve cells called "whisker barrels" in their brains, because their whiskers are sense organs. I've got a mustache, but I don't have whisker barrels because my mustache is not a very important sensory organ for me.

So if you see this fetal mouse making normal whisker barrels, you feel comfortable. If it fails to make whisker barrels, that's all right too. But if you see no whisker barrels forming and you see a human visual cortex forming there instead, if you see something weird, stop the experiment and talk. If you see nothing weird, let them be born, watch both their brains and behavior, and if you see something weird, stop them.

MR. SAFIRE: Talk about a slippery slope. We're not talking about mixing a zebra with a horse or a lion with a tiger, which has happened. We're talking about mixing the human brain with an animal brain, and a lot of people think that is, to use the analogy of the Greek, monstrous. I was reading in your book, Mike, about a word I had not seen before, a "humanzee," which I presume is a combination of "chimpanzee" and "human."

DR. GAZZANIGA: If this chimeric problem had been presented to me, I would have said that neuroscience is in no way able to identify the neural circuits, the very organization that underlies the production of any sort of conscious behavior that is humanlike. It's not even able to work out the circuits that are involved in seeing a triangle, let alone one of the higher cognitive functions. So the probability that a few neurons injected into a mouse would generate a little Harry in there is so without meaning, given modern knowledge of neuroscience, that I would have simply viewed the mouse as a convenient tissue culture for the growth of these stem cells to see how they would develop and what would happen.

MR. GREELY: There is an institutional problem with the field of bioethics. There are some rewards in terms of publicity and attention for being more extreme and more dramatic, but I think most of us who work in that field are more responsible and try not to raise outrageous examples simply for the purpose of getting discussion going, except in an undergraduate seminar, where outrageous examples are useful.

DR. GAZZANIGA: Agreed.

MR. SAFIRE: Or a Library of Congress discussion.

(Laughter)

MR. SAFIRE: Let us now turn to a controversy that we were just talking about among ourselves. Enhancement versus treatment—that's a big ethical issue. Let me pose it this way: It's a good idea to treat failing memory in aging adults, but what would you say to providing students a pill to enhance memory just before an academic exam? Would that be akin to giving athletes steroids?

MR. GREELY: I think it might be akin to giving athletes steroids. I think whether athletes taking steroids is a bad thing or not is much more complicated than has generally been considered. It is a way to enhance people. Is it a good way or a bad way? As for the memory pill, it would depend on a lot of things for me. You're asking a lawyer, so the answer is always "It depends," and then I send you the bill. But it really would depend on whether it was safe, whether it worked for longer than five hours, and so on. If all it does is get the students through the test and then they forget the CREB cycle and have to memorize it again in the next step of their pre-med curriculum, it's different. Does everybody in the class get this pill, or only a few people?

All those, I think, are relevant questions about it, but enhancement in general shouldn't be a dirty word. I'm a teacher; enhancement is my business. I try to enhance people, both by giving them knowledge and by giving them ways to manipulate knowledge, and that actually changes the physical insides of their brains.

DR. GAZZANIGA: Well, I think the enhancement issue should be broken down into at least two areas. Let's deal first with the competitive sports metaphor, and then later with the mental pill, which I could use right now.

(Laughter)

DR. GAZZANIGA: When you're an athlete, you've cut a deal. The deal is that the competing teams are using the same food source and pharmacological sources, or lack of them. For one team to sneak in and to pump themselves up on drug X is cheating. That's not part of the sport. The sport is to go in there and have the social contract, let the athlete value the hard work of practice and all the rest of it. So I'm very much against letting drugs slip into the athletic arena, unless you want to go into the professional arena, where it becomes more of an entertainment issue and less of an athletic issue.

But when it comes to pharmaceutical aids for memory, I think we have to keep in mind that there you're monkeying with your own ability to encode information at a faster rate, and if you, Bill Safire, want to do that, it's fine with me. You must remember that the drugs that we're talking about are not in any way going to increase your intelligence. They're just going to increase the rate with which you apprehend the new information to think about. So I think these are two different arenas.

MR. GREELY: I don't know how different they really are, but I agree that I'm not going to endorse cheating. But the question really for me becomes, What should the rules be? Should we have rules against pharmacological enhancements, either for sports or for academics? If there are serious safety risks, that's one good argument for a rule against it, if you think you can enforce it successfully. If there are fairness issues, where one side gets it and another side doesn't, that's another reason to think about regulation—maybe banning it, maybe making sure everybody gets it. But I don't know that there's any inherent reason why enhancement is a dirty word and why using enhancement pharmacologically is any different from spending time at the weight room, having your diet carefully regulated, going to a sports psychologist, going to the best coach.

MR. SAFIRE: A lot of people feel that enhancement carried to its extreme undermines our humanity.

MR. GREELY: It may be that enhancement carried to its extremes is wrong, but then it's wrong to carry it to its extremes. Then we have to get into the argument about where we define the extremes.

MR. SAFIRE: Okay. Now, you're a lawyer. It's important for the justice system to make sure that nobody bears false witness. Is it right ethically to image a brain to detect perjury?

MR. GREELY: You know, there are at least four different groups working on it and four different techniques. Paul Wolpe here in the audience has an article coming out about various neuroscience lie detectors. It may never work. Most likely, it'll work a little but not a lot. But assuming it did work, I don't think it would start with somebody saying, "I object to that testimony and let's put them under the lie detector." I think it would start with individuals saying, "I'm telling the truth, and here's the lie detector test that proves it." Then the courts would have to answer the question raised by somebody who volunteers to submit to a lie detector test to prove, however solid the proof is, that they're telling the truth: Should the courts allow it?

If the technology turns out to be truly reliable, I think our courts would have difficulty saying no. Whether they'll force people to take it is a harder question and involves a lot of constitutional issues, but allowing people to take it I think is likely to happen.

MR. SAFIRE: Mike, how do you feel about the invasion of privacy if the day comes when you can actually see the amygdala light up and say, "That means he's lying"?

DR. GAZZANIGA: The activation pattern in a brain scan doesn't tell you the psychological content of the person's thoughts. So I see the probability that we will have a foolproof lie-detecting system as remote. That's point one.

Now, point two has to do with the individual variation in brain scanning, which is now finally being examined by many, many laboratories. We've grown up in the last fifteen years thinking when we see one of these scans that they finally have figured out the brain circuits involved in how you do whatever the problem is you're working on. But it turns out those are averages and that the individual variations are huge. One of the goals here is to imagine the neuroscientist comes into the courtroom and says, "Look, this pixel in the brain here is lit up and so therefore Harry didn't kill the person." But the other side will simply have to point to the variation in the response, and I think once the extent of variation is seen by the court, that kind of claim couldn't be made.

MR. SAFIRE: Let's assume you can, in years to come, predict future behavior through a combination of brain scanning and genetic study. Should insurance companies be informed that so-and-so has a disease that's liable to come out in twenty years? Should the police be informed that so-and-so has a real inclination to violence that might turn to murder in a few years?

MR. GREELY: I really do think it depends on the circumstances. If there's an intervention that the insurance company can pay for that can stop you from developing the disease, sure, let the insurance company know, let them pay for the intervention. If it is something where the insurance company is only interested in medical underwriting and getting you off their rolls because it's an expensive disease, well, I'm opposed to that. We could solve this whole problem by joining every other rich country and having guaranteed health care for all our people, then we wouldn't have that problem, but as long as we aren't there, I'm not in favor of additional ways to keep people off health insurance.

The police example, I think, is harder. It depends in large part on just how good your information is. Have you done a 10,000-person trial that shows that 99.99 percent of the time this signal is followed by significant violent activity? Even then, it depends on what your intervention would be. Do you throw somebody in jail before they've ever done anything that you can prove? I don't think so, but do you put them into some sort of diversion program to try to help them work out their anger, or do you tell the neighbors, or do you tell people to watch out for them? The kind of intervention, as well as the strength of your prediction, is very significant in whether this would be a good thing or a bad thing.

MR. SAFIRE: Mike, where do you come out on this?

DR. GAZZANIGA: We're in just an interim position and an impossible position. The flow of information and knowledge about people's health state is going to be known and it's going to be known quickly and completely. So from a theoretical point of view, given where biology is going, I think our political system simply has to accept the fact that we need a one-payer health system.

MR. SAFIRE: Put yourself in the shoes of a doctor or a scientist who has the scientific means to predict mental or behavioral problems for which they have no effective treatment. Do you tell the patient about it, or do you protect him from the knowledge of his own future disability?

MR. GREELY: I think you ask him or her whether he or she wants to know and let them know in advance what some of the benefits and costs of knowledge would be and let that person make his or her own decision. We need a third-person neuter pronoun, don't we?

MR. SAFIRE: That's another subject.

MR. GREELY: Let those people make their own decision based on what's right for them. The thing that is most parallel between genetics and neuroscience is these prediction issues, because they raise the same kinds of questions about knowledge that you can't act on. There's a nice crossover example: Huntington's disease is both a genetic and a neurological disease. If you've got the variation of the gene that causes Huntington's disease, as far as we know, the only way you can avoid getting the disease is to die first from something else. It's a nasty disease with no treatment, no prevention, and it usually strikes people in midlife, in their 30s through 50s.

We've been able to test for the Huntington variation of this gene for about a decade and a half now. Interestingly, in the U.S., only about 30 percent of the people who are at risk and who make contact with the testing center actually decide to get tested—and half of those who get tested don't pick up their test results. But I say as long as they're competent adults, you let people make the decision for themselves about whether they want this information.

MR. SAFIRE: Okay. What about an employer who hires an outfit to go to for a screening and this outfit knows everything about the bank account and the academic record, and so on? Do you then allow the kind of screening we're talking about, about proclivities toward violence or sexual predation or embezzlement?

MR. GREELY: I think for me the answer would be almost never, but if it is something that has a strong health or safety risk for this particular employee with this particular predisposition to be in this particular job—in that very limited case, I'd be willing to make an exception along the lines of something that the Americans with Disabilities Act actually has, which is a health and safety exception.

MR. SAFIRE: How about a candidate for president?

MR. GREELY: Well, you're the history buff. How many of our past presidents had some significant health or other issues they were hiding? Half of them?

MR. SAFIRE: I knew he'd drag my Nixon background in here.

MR. GREELY: I don't think we should force candidates for office to reveal their medical records, but I think we're within our rights to ask them and to take that into consideration in deciding how to vote.

DR. GAZZANIGA: Well, predicting the future is a bum's game; no one ever gets it right. If you look at the probability that identical twins will both have a disease of a particular kind, it's hard to predict. If you look at the variation in their behavior, it has to do with their individual experiences. It has to do with the fateful chance of different experience as they develop, given that they have a tremendous amount in common in their bodies and their mental processes.

MR. GREELY: I disagree. It's easy to predict; it's just hard to be right. Powerful accurate predictions are going to be rare, but there are some, like Huntington's allele corresponding to Huntington's disease—that's a pretty powerful prediction. But you don't need real good predictions if you're dealing with large numbers. It's important for us to remember how often our society uses predictions. I've got a sixteen-year-old son. He just got his driver's license. I just got my insurance bill.

(Laughter)

MR. GREELY: Here's another example: university admissions offices predict future success based on the SATs, grades, recommendations, legacy status, a variety of different things, some of which are reasonable predictors and some of which are bad predictors, but taken all in all, they're better than nothing. If you're really trying to gamble, you don't need to be right every time. You only need to be right 52 percent of the time to make money.

MR. SAFIRE: But coming back to the idea of enhancement, does it trouble you at all that we're making life fat, sassy, and easy for people, for young people particularly, by giving them drugs that will prevent them from becoming sad or prevent them from losing their attention? Doesn't this undermine the natural development of the human being to become self-reliant and to overcome obstacles?

DR. GAZZANIGA: There definitely are these interfering drugs. Randolph Nesse wrote a book called *Why We Get Sick,* and the important point of the book is that our emotional system has all kinds of important cuing mechanisms to tell us when to stop overreasoning on the continuum between sadness and depression. That's a good thing to have. If you intervene with a drug that might clip that, you might be not quite making the right sort of social judgments as you go through your life. There are real issues that call for very careful thought about how these drugs have unintended consequences with your decision-making system.

MR. GREELY: Sometimes I do feel that life is too easy for young people today, but, first, there are a lot of kids in this country for whom life is not easy at all. A shockingly high percentage of the children in the United States are in poverty and are in difficult situations, let alone the children of the world. Even when privileged children avoid those traumas, there are plenty of other traumas in the world to be had, from the death of a loved one to accidents to disease, and so on.

MR. SAFIRE: I'm not saying make it harder for them. I'm just saying don't make it easier for them.

MR. GREELY: Don't let them be better.

MR. SAFIRE: Don't let them enhance themselves.

MR. GREELY: Because it would take them away from the state of nature in which life is brutish, nasty, and short, right?

MR. SAFIRE (to Dr. Gazzaniga): Let's go to the big question in your book. Is there an inherent moral sense, or is this something that we develop and are taught, not to murder and not to steal and not to break these commandments?

DR. GAZZANIGA: This has a three-part answer. I've spent my life studying a system in the left hemisphere of our brains that tends to interpret the behaviors in our felt states of emotion. We build a theory about why we feel a particular way or why we did a particular thing, and that becomes our narrative, our personal story. These narratives exist and we know where they are in the brain, and all kinds of studies have shown that. So that's point one.

Now, for point two, here is an experiment: Think that you're the only person in the world. Everybody else doesn't exist. You're it. Now, write down everything that you've been thinking about in the last forty-eight hours. When you do that little armchair experiment, you will notice that 99 percent of the things you've been thinking about have to do with other people. All we do all day long is think about the intentionality of other people, social comparisons of where we are, the variety of complex social processes that you and I are involved in. Somehow our species is pretty good at getting along, and so we don't kill, we don't like to cheat, we don't like incest. The idea is there's some kind of moral compass.

Point three: Maybe as a species, all of us tend to react in the same way to certain moral dilemmas. There's been an explosion of exciting new work on this very point: we all tend to respond throughout the world, throughout the culture, throughout religious belief systems, throughout age, the same way. Why do we do that? Maybe it turns out there are brain circuits involved that bring us back from the brink of responding in a particular way, bring us back from the brink of killing or from the brink of cheating more often than not, and that just happens. We just do it this way. We stop, and then because we stop, our little interpreter builds up a theory as to why we do that, and we have a narrative for ourselves and for our culture.

That's a long-winded way of saying that I think the excitement of neurobiology in the next twenty years is going to be to figure out what underlies these kinds of social processes, what are the circuits that are involved in keeping our species not at each other's throats. People forget that there have been only five thousand generations between everybody now in the world and the time when there were only ten thousand people on earth. We all have the same genes and that instinctual response. I think modern neuroscience is going to begin to pick those circuits out, show that they're real, and explain why we behave the way we behave. Each of us then spins a different story, or cultures spin different stories, as to why, and that's fine, but it's secondary to the primary built-in common circuitry.

MR. SAFIRE: Now, Hank, just following that out, if indeed there is a moral sense, how does that impact on the courtroom, on a person's responsibility over right and wrong and claiming insanity or claiming no responsibility?

MR. GREELY: I guess this is an issue that philosophers and a lot of lawyers are interested in: free will and its effect with respect to criminal liability. If you can prove that I didn't have a choice, that my anterior cingulate cortex made me do it, would the jury or would the judge let me off? For the most part, juries haven't been terribly sympathetic and neither have judges, but occasionally it works. My own sense is that even if a neuroscientist could prove to us all that there was no such thing as free will, we would ignore him in the criminal setting. We would continue to treat people as if they are responsible, whether we actually believe they are or not, because it's so deeply ingrained in our culture and it seems plausible, at least, that it's a useful thing to do in terms of deterrent and other effects. I'd be really shocked if our culture were to give it up.

DR. GAZZANIGA: You have to distinguish between free will and personal responsibility. Personal responsibility is a social construct. If you're the only person in the world, there's no concept of personal responsibility. Who are you personally responsible to? Stone? The ants? I think where neuroscientists are going to make a mistake is if they get dragged into a courtroom and say, "There's reduced personal responsibility because of this brain scan." That's not where you find personal responsibility. It's a rule of the social group; it's not to be found in the brain. As I like to say, schizophrenics stop at red lights, you know; they follow rules, and rules can be followed in all kinds of diminished states. I would be very reluctant to give up the concept of personal responsibility in any kind of legal setting.

MR. GREELY: Even if you were to give up the concept of free will?

DR. GAZZANIGA: All free will means is that as you become an educated person, your brain has more options to choose from. So you're freer to have a larger range of options. But all of neuroscience is driving towards a mechanistic understanding of us, and, with every passing year, we are finding how automatic these processes are that we used to hold dear as somehow not automatic. But that doesn't lessen the social contract that emerges when people come together as a group and they write down those rules.

MR. SAFIRE (to audience): If this discussion has stimulated some questions in your mind, or if you've been waiting to ask one that I haven't gotten near, feel free to raise your hand.

PARTICIPANT: I was wondering how the advances in neurobiology would affect the political setting in terms of homosexuality, like if there's a cure for it, because some research indicates that it is a neurological disease, while some research indicates that it may be something environmentally stimulated, and I was wondering how that would affect neuroethics.

MR. GREELY: It's really hard to know. It's hard to know which direction any of these arguments would cut. The research I think you're referring to came out actually in *Genetics* about ten years ago. A scientist at NIH published a finding that he said showed there was a gene mutation that predisposed strongly toward male homosexuality. I think that finding is now generally not accepted to have been correct, but the research article actually prompted a significant debate about whether, if it were true, it would be a good thing or a bad thing for the gay community. And the answer, depending on who you talked to, was yes, no, and maybe, because even if you make the logical jump from "genetic" to "unavoidable" (not necessarily an accurate jump), the argument about unavoidability could cut in both directions. It could be liberating, or it could be very tyrannizing. So if there is further information, either about the source of nonmajoritarian sexual preferences or about ways to change them, I wouldn't venture a prediction about how it would play out in society—although I will predict that it'll play out differently in San Francisco than it will in Topeka and differently in both than it will in Riyadh or Tehran.

MR. SAFIRE: We have a question from a gentleman whose organization really is the pioneer over the last twenty-five years of bioethics. Let's hear from you.

DR. MURRAY: Hank had some fairly blasé comments about the ethics of enhancement. But let's say we do discover neurocircuits. Say one of them allows us to feel your pain and really feel it—not just recognize it but feel it. And let's say that I then came to you as my doctor and said, "I'd like you to enhance me by tamping down my ability to feel other people's pain. I want to be able to recognize it because I want to be able to manipulate other people, but I think this actually will stand me very good in my chosen field of . . ." and you can fill in the blank, you can make it politics, litigation, whatever—and this, to me, would be an enhancement. I could practice my craft more successfully and with less discomfort than I could otherwise. Now, you were blasé about enhancement. Are you blasé about this enhancement?

MR. GREELY: I don't think it's fair to say that I'm blasé about it, although I am not universally opposed to it. I think the answer is it depends, and in the case that you've given me, the answer again is, it depends. It depends in part on how safe the procedure is; what the side effects are; how realistic I think this person's assessment is of its benefits and costs for him; and then finally, as a physician, if you make me a physician, whether this is the sort of business I want to be in, whether this is something I choose to spend my time and effort on rather than doing treatment. All of those things would factor into it, but I will not say, as a general matter, that it would be wrong. I don't think tamping down traumatic experiences that people have gone through is necessarily a bad thing.

PARTICIPANT: Some years ago, I heard a talk on prescribed medication and the requirement for people to be forced to take their medication. How would this be a requirement, and what would be its effect on society? Can you require people to take their medication, and if they don't go along, is it an infringement of their civil liberties if you're forcing them to take it?

DR. GAZZANIGA: There are famous court cases on this, including one just adjudicated, in which the question was whether to force the defendant to take a drug to keep him sane enough during the course of the trial so that he could then be convicted of the crime involved. This is a hotly debated issue. It's easy to think about these things in a seminar room, but then how is it actually going to work in the real world?

I'd like to leave maybe on one note that I think sums up what we're trying to do here today and speaks to the problems that are out there. It was just a little over a year ago that one of the speakers at our Bioethics Council meeting was looking at the question not of memory enhancement but the flip side, memory erasure.

MR. SAFIRE: This has to do with the President's Council on Bioethics, on which Mike serves and is usually a dissenting voice.

DR. GAZZANIGA: The two scientists that were involved were leaders in the field, and their notion is as follows. You take an animal or a person. You bring up into their mind's eye the memory of a past experience; then you zap them, and you can do away with the memory. People were fearful of what this procedure would do to the personality structure, but overall it was dismissed as not being really hard neuroscience and probably not true.

Two weeks ago, at the Cognitive Neuroscience Society, two thousand people were sitting and listening to a full three-hour symposium in which there were sixty experiments on this topic. Yes, it is true; yes, you can do it. And yes, some researchers are already thinking of pills that you could have, such that if you're having or about to have a traumatic experience, like meeting someone you don't like or an automobile accident in which you get beaten up, you take this pill and the memory for the event will be so knocked down that you will remember that it happened, but it won't be the kind of emotionally intensive memory that leads to a posttraumatic syndrome. So you can imagine soldiers with these pills, and so forth. My point is that in one year, we've gone from "This can't possibly be true" to the fact that it is true. So I think that groups such as this, who worry about bioethical issues, have some real serious questions to think about as to how they want to handle such a drug in society. I think those things abound; they're everywhere. They're seeping out of brain science, and I am, and I'm sure Hank is also, grateful that you give us a chance to shoot our mouths off about it.

MR. SAFIRE: And with that, thanks very much for coming.

Further Reading

Angold, A., et al. "Stimulant Treatment for Children: A Community Perspective." *Journal of the American Academy of Child and Adolescent Psychiatry* 39, no. 8 (2000): 975–984.

Ashby, P., et al. "Immediate Motor Effects of Stimulation through Electrodes Implanted in the Human Globus Pallidus." *Stereotactic and Functional Neurosurgery* 70, no. 1 (1998): 1–18.

Bechara, A., D. Tranel, and H. Damasio. "Characterization of the Decision-Making Deficit of Patients with Ventromedial Prefrontal Cortex Lesions." *Brain* 123, no. 11 (2000): 2189–2202.

Friehs, G. M., et al. "Brain-Machine and Brain-Computer Interfaces." *Stroke* 35, no. 11, suppl. 1 (2004): 2702–2705.

Greene, J., et al. "The Neural Bases of Cognitive Conflict and Control in Moral Judgment." *Neuron* 44, no. 2 (2004): 389–400.

Jackson, P. L., A. N. Meltzoff, and J. Decety. "How Do We Perceive the Pain of Others? A Window into the Neural Processes Involved in Empathy." *Neuroimage* 24, no. 3 (2005): 771–779.

McGaugh, J. "Remembering and Forgetting: Physiological and Pharmacological Aspects." Transcript of a presentation at a meeting of the President's Council on Bioethics, Washington, D.C., 17 October 2002 (www.bioethics.gov/transcripts/oct02/session3.html).

Macklin, R. "Some Questionable Premises about Research Ethics." *American Journal of Bioethics* 5, no. 1 (2005): 29–31.

Malone, K. M., et al. "Functional Brain Magnetic Resonance Imaging (fMRI) Study of Problem-Solving and Hope in Healthy Volunteers." Paper presented at the 43rd annual meeting of the New Clinical Drug Evaluation Unit (NCDEU), Boca Raton, Fla., 2003.

Marsh, E. J., K. B. McDermott, and H. L. Roediger. "Does Test-Induced Priming Play a Role in the Creation of False Memories?" *Memory* 12, no. 1 (2004): 44–55.

Mayberg, H. S., et al. "Deep Brain Stimulation for Treatment-Resistant Depression." *Neuron* 45, no. 5 (2005): 651–660.

Meyer-Lindenberg, A., et al. "Midbrain Dopamine and Prefrontal Function in Humans: Interaction and Modulation by COMT Genotype." *Nature Neuroscience* 8, no. 5 (2005): 594–596.

Murray, H. *Explorations in Personality: A Clinical and Experimental Study of Fifty Men of College Age.* New York: John Wiley & Sons, 1938.

National Institute on Drug Abuse. "Teen Drug Use Declines 2003–2004— But Concerns Remain About Inhalants and Painkillers." News release, 21 December 2004 (www.drugabuse.gov/Newsroom/04/ NR12-21.html).

Schacter, D. L. "Remembering and Forgetting: Psychological Aspects." Transcript of a presentation at a meeting of the President's Council on Bioethics, Washington, D.C., 17 October 2002 (www.bioethics.gov/ transcripts/oct02/session4.html).

Schiff, N. D., et al. "fMRI Reveals Large-Scale Network Activation in Minimally Conscious Patients." *Neurology* 64, no. 3 (2005): 514–523.

Sitton, M., M. C. Mozer, and M. J. Farah. "Superadditive Effects of Multiple Lesions in a Connectionist Architecture: Implications for the Neuropsychology of Optic Aphasia." *Psychological Review* 107, no. 4 (2000): 709–734.

Slotnick, S. D., and D. L. Schacter. "A Sensory Signature That Distinguishes True from False Memories." *Nature Neuroscience* 7, no. 6 (2004): 664–672.

Zuvekas, S.H., Vitiello, B., and Norquist, N.S. "Recent Trends in Stimulant Use Among U.S. Children." *American Journal of Psychiatry* (forthcoming 2006).

Weizenbaum, J. *Computer Power and Human Reason: From Judgment to Calculation.* San Francisco: W. H. Freeman, 1976.

Wolpe, P. R., K. R. Foster, and D. D. Langleben. "Emerging Neurotechnologies for Lie-Detection: Promises and Perils." *American Journal of Bioethics* 5, no. 2 (2005): 39–49.

World Health Organization. *The World Health Report 2001. Mental Health: New Understanding, New Hope.* Geneva: World Health Organization, 2001.

Index

Other Dana Press Books And Periodicals

www.dana.org/books/press

BOOKS FOR GENERAL READERS

Brain and Mind

THE CREATING BRAIN: *The Neuroscience of Genius*

Nancy C. Andreasen, Ph.D., M.D.

Andreasen, a noted psychiatrist and bestselling author, explores how the brain achieves creative breakthroughs—in art, literature, and science—including questions such as how creative people are different and the difference between genius and intelligence. She also describes how to nurture and develop our creative capacity. 33 illustrations/photos. 225 pp.

1-932594-07-8 • $23.95
eBook 1-932594-18-3 • $12

THE ETHICAL BRAIN

Michael S. Gazzaniga, Ph.D.

Explores how the lessons of neuroscience help resolve today's ethical dilemmas, ranging from when life begins to "off-label" use of drugs such as Ritalin by students preparing for exams, and other topics, from free will and personal responsibility to public policy and religious belief. The author, a pioneer in cognitive neuroscience, is a member of the President's Council on Bioethics. 225 pp.

1-932594-01-9 • $25.00
eBook 1-932594-21-3 • $12

A GOOD START IN LIFE: Understanding Your Child's Brain and Behavior from Birth to Age 6

Norbert Herschkowitz, M.D., and Elinore Chapman Herschkowitz

Updated with the latest information and new material, the authors show how young children learn to live together in family and society and how brain development shapes a child's personality and behavior, discussing appropriate rule-setting, the child's moral sense, temperament, language, playing, aggression, impulse control, and empathy.

Cloth 283 pp. 0-309-07639-0 • $22.95
Paper (Updated version with 13 illustrations) 312 pp.
0-9723830-5-0 • $13.95

BACK FROM THE BRINK: How Crises Spur Doctors to New Discoveries about the Brain

Edward J. Sylvester

In two academic medical centers, Columbia's New York Presbyterian and Johns Hopkins Medical Institutions, a new breed of doctor, the neurointensivist, saves patients with life-threatening brain injuries. 16 illustrations/photos. 296 pp.

0-9723830-4-2 • $25.00

THE BARD ON THE BRAIN: Understanding the Mind Through the Art of Shakespeare and the Science of Brain Imaging

Paul Matthews, M.D., and Jeffrey McQuain, Ph.D.
Foreword by Diane Ackerman

Explores the beauty and mystery of the human mind and the workings of the brain, following the path the Bard pointed out in 35 of the most famous speeches from his plays. 100 illustrations. 248 pp.

0-9723830-2-6 • $35.00

STRIKING BACK AT STROKE: A Doctor-Patient Journal

Cleo Hutton and Louis R. Caplan, M.D.

A personal account with medical guidance for anyone enduring the changes that a stroke can bring to a life, a family, and a sense of self. 15 illustrations. 240 pp.

0-9723830-1-8 • $27.00

THE DANA GUIDE TO BRAIN HEALTH

Floyd E. Bloom, M.D., M. Flint Beal, M.D., and David J. Kupfer, M.D., Editors. Foreword by William Safire

A home reference on the brain edited by three leading experts collaborating with 104 distinguished scientists and medical professionals. In easy-to-understand language with cross-references and advice on 72 conditions such as autism, Alzheimer's disease, multiple sclerosis, depression, and Parkinson's disease. 16 full-color pages and more than 200 black-and-white illustrations.

Cloth 768 pp. 0-7432-0397-6 • $45.00

UNDERSTANDING DEPRESSION: What We Know and What You Can Do About It

J. Raymond DePaulo Jr., M.D., and Leslie Alan Horvitz. Foreword by Kay Redfield Jamison, Ph.D.

What depression is, who gets it and why, what happens in the brain, troubles that come with the illness, and the treatments that work.

Cloth 304 pp. 0-471-39552-8 • $24.95
Paper 296 pp. 0-471-43030-7 • $14.95

KEEP YOUR BRAIN YOUNG: The Complete Guide
to Physical and Emotional Health and Longevity

Guy McKhann, M.D., and Marilyn Albert, Ph.D.

Every aspect of aging and the brain: changes in memory, nutrition, mood, sleep, and sex, as well as the later problems in alcohol use, vision, hearing, movement, and balance.

Cloth 304 pp. 0-471-40792-5 • $24.95
Paper 304 pp. 0-471-43028-5 • $15.95

THE END OF STRESS AS WE KNOW IT

Bruce McEwen, Ph.D., with Elizabeth Norton Lasley.
Foreword by Robert Sapolsky

How brain and body work under stress and how it is possible to avoid its debilitating effects.

Cloth 239 pp. 0-309-07640-4 • $27.95
Paper 262 pp. 0-309-09121-7 • $19.95

IN SEARCH OF THE LOST CORD: Solving the Mystery of
Spinal Cord Regeneration

Luba Vikhanski

The story of the scientists and science involved in the international scientific race to find ways to repair the damaged spinal cord and restore movement. 21 photos; 12 illustrations. 269 pp.

0-309-07437-1 • $27.95

THE SECRET LIFE OF THE BRAIN

Richard Restak, M.D. Foreword by David Grubin

Companion book to the PBS series of the same name, exploring recent discoveries about the brain from infancy through old age. 201 pp.

0-309-07435-5 • $35.00

THE LONGEVITY STRATEGY: How to Live to 100 Using the Brain-Body Connection

David Mahoney and Richard Restak, M.D.
Foreword by William Safire

Advice on the brain and aging well.

Cloth 250 pp. 0-471-24867-3 • $22.95
Paper 272 pp. 0-471-32794-8 • $14.95

STATES OF MIND: New Discoveries About How Our Brains Make Us Who We Are

Roberta Conlan, Editor

Adapted from the Dana/Smithsonian Associates lecture series by eight of the country's top brain scientists, including the 2000 Nobel laureate in medicine, Eric Kandel.

Cloth 214 pp. 0-471-29963-4 • $24.95
Paper 224 pp. 0-471-39973-6 • $18.95

Immunology

FATAL SEQUENCE: The Killer Within

Kevin J. Tracey, M.D.

An easily understood account of the spiral of sepsis, a sometimes fatal crisis that most often affects patients fighting off nonfatal illnesses or injury. Tracey puts the scientific and medical story of sepsis in the context of his battle to save a burned baby, a sensitive telling of cutting-edge science. 225 pp.

Cloth 1-932594-06-X • $23.95
Paper 1-932594-09-4 • $12.95
eBook 1-932594-14-0 • $9.99

Arts Education

A WELL-TEMPERED MIND: Using Music to Help
Children Listen and Learn

Peter Perret and Janet Fox. Foreword by Maya Angelou

Five musicians enter elementary school classrooms, helping children learn about music and contributing both to higher enthusiasm and improved academic performance. This charming story gives us a taste of things to come in one of the newest areas of brain research: the effect of music on the brain. 12 illustrations. 225 pp.

Cloth 1-932594-03-5 • $22.95
Paper 1-932594-08-6 • $12.00
eBook 1-932594-20-5 • $9.00

THE DANA FOUNDATION SERIES ON NEUROETHICS

NEUROSCIENCE AND THE LAW: Brain, Mind, and the Scales of Justice

Brent Garland, Editor. Foreword by Mark S. Frankel. With commissioned papers by Michael S. Gazzaniga, Ph.D., and Megan S. Steven; Laurence R. Tancredi, M.D., J.D.; Henry T. Greely, J.D.; and Stephen J. Morse, J.D., Ph.D.

How discoveries in neuroscience influence criminal and civil justice, based on an invitational meeting of 26 top neuroscientists, legal scholars, attorneys, and state and federal judges convened by the Dana Foundation and the American Association for the Advancement of Science. 226 pp.

1-932594-04-3 • $8.95

BEYOND THERAPY: Biotechnology and the Pursuit of Happiness.

A Report of the President's Council on Bioethics

Special Foreword by Leon R. Kass, M.D., Chairman.
Introduction by William Safire

Can biotechnology satisfy human desires for better children, superior performance, ageless bodies, and happy souls? This report says these possibilities present us with profound ethical challenges and choices. Includes dissenting commentary by scientist members of the Council. 376 pp.

1-932594-05-1 • $10.95

NEUROETHICS: Mapping the Field. Conference Proceedings.

Steven J. Marcus, Editor

Proceedings of the landmark 2002 conference organized by Stanford University and the University of California, San Francisco, at which more than 150 neuroscientists, bioethicists, psychiatrists and psychologists, philosophers, and professors of law and public policy debated the implications of neuroscience research findings for individual and societal decision-making. 50 illustrations. 367 pp.

0-9723830-0-X • $10.95

FREE EDUCATIONAL BOOKS

(Information about ordering and downloadable PDFs are available at www.dana.org.)

PARTNERING ARTS EDUCATION: A Working Model from ArtsConnection

This publication illustrates the importance of classroom teachers and artists learning to form partnerships as they build successful residencies in schools. *Partnering Arts Education* provides insight and concrete steps in the ArtsConnection model. 55 pp.

ACTS OF ACHIEVEMENT: The Role of Performing Arts Centers in Education

Profiles of more than 60 programs, plus eight extended case studies, from urban and rural communities across the United States, illustrating different approaches to performing arts education programs in school settings. Black-and-white photos throughout. 164 pp.

PLANNING AN ARTS-CENTERED SCHOOL: A Handbook

A practical guide for those interested in creating, maintaining, or upgrading arts-centered schools. Includes curriculum and development, governance, funding, assessment, and community participation. Black-and-white photos throughout. 164 pp.

THE DANA SOURCEBOOK OF BRAIN SCIENCE: Resources for Teachers and Students, Fourth Edition

A basic introduction to brain science, its history, current understanding of the brain, new developments, and future directions. 16 color photos; 29 black-and-white photos; 26 black-and-white illustrations. 160 pp.

THE DANA SOURCEBOOK OF IMMUNOLOGY: Resources for Secondary and Post-Secondary Teachers and Students

An introduction to how the immune system protects us, what happens when it breaks down, the diseases that threaten it, and the unique relationship between the immune system and the brain. 5 color photos; 36 black-and-white photos; 11 black-and-white illustrations. 116 pp.

PERIODICALS

Dana Press also offers several periodicals dealing with arts education, immunology, and brain science. These periodicals are available free to subscribers by mail. Please visit www.dana.org.